Emory Pratt
11-17-84
27.95

Ecology of Tropical Plants

Ecology of Tropical Plants

Margaret L. Vickery

with a chapter by
Dr. John Hall
University of Dar es Salaam

JOHN WILEY & SONS
Chichester · New York · Brisbane · Toronto · Singapore

Parts of this book have been adapted from
'Plants and Environment'
by R. F. Daubenmire. 3rd Edition

Library of Congress Cataloging in Publication Data:

Vickery, Margaret L.
Ecology of tropical plants.

'Adapted from Plants and environment, a textbook of autecology, 3rd edition, by R. F. Daubenmire' – T.p. verso.
Includes index.
1. Tropical plants – Ecology. 2. Botany – Tropics – Ecology. I. Daubenmire, Rexford F., 1909– . Plants and environment. 3rd. ed. II. Title.
QK936.V53 1984 581.5′2623 83-6973
ISBN 0 471 90107 5 (cloth)
ISBN 0 471 90200 4 (paper)

British Library Cataloguing in Publication Data:

Vickery, Margaret
Ecology of tropical plants.
1. Tropical plants
I. Title II. Hall, John
591.909′3 QK936

ISBN 0 471 90107 5 (cloth)

Typeset by Oxford Verbatim Limited
Printed in Great Britain by
Pitman Press Limited, Bath, Avon

Contents List

Preface

This textbook is an adaptation, for tropical regions, of the established book by R. F. Daubenmire, *Plants and Environment – A textbook of Autecology*, 3rd edition, John Wiley and Sons Inc., 1974. The arrangement follows that of Daubenmire in that each facet of the physical and organic environment of plants is treated separately, starting with soil and ending with the influence of man. However, the text concentrates exclusively on the tropical environment and its influence on plants. The book contains much new and updated material, including a chapter devoted to practical aspects of the subject contributed by Dr. J. Hall.

Ecology is now firmly established as a discipline in its own right and its importance to the future well-being of mankind is fully recognized. Tropical ecosystems are both complex and highly vulnerable to interference by man. It is only through the study of such systems by ecologists and biogeographers that the effects of this interference can be predicted and, hopefully, steps taken to minimize the irreversible destruction which is now taking place in many areas of the tropics. During my ten-year residence in tropical Africa I was able to witness at first hand the disastrous effects on the environment of forest clearance, overcultivation, and overgrazing.

The tropical region encompasses four continents – Asia, Africa, South America, and Australasia. Although many environmental characteristics are common to all tropical regions, there are variations. This is especially true of the occurrence of individual plant species. While some are pantropical, others are much more restricted in their distribution. Thus in this book examples of individual species have been kept to a minimum and only quoted when some specific point needed illustrating. It is hoped that lecturers in the tropics using this book will expand the text by including examples of local plant species.

Finally, I would like to thank Dr. Hall for his contribution, 'Investigating the Environment'. This is a most valuable addition to the book, as it is only by carrying out field work that the principles of ecology can be fully appreciated by students.

Margaret L. Vickery

Chapter 1

Introduction

Organisms cannot live in isolation and the study of their continual interaction with their environment, known as *ecology*, is an important facet of biology. The term *ecology* is derived from the Greek word for house and translates as the study of all members of a household. If *habitat* is substituted for household, then ecology can be scientifically defined as the study of habitats.

All organisms are affected by their environment, both living (other organisms) and non-living (soil, climate, etc.), the total influence of these environmental factors making up the habitat in which the organism lives. For animals that move around, a habitat can be a large area (for man it can cover the entire world), but for plants habitats can be quite small.

The basic functional unit of ecology is the *ecosystem*. An ecosystem is a functional, interacting entity which includes all the living organisms, termed a *community*, and the non-living environment of a particular area (fig. 1.1). All the earth's ecosystems taken together make up the *biosphere* (fig. 1.2).

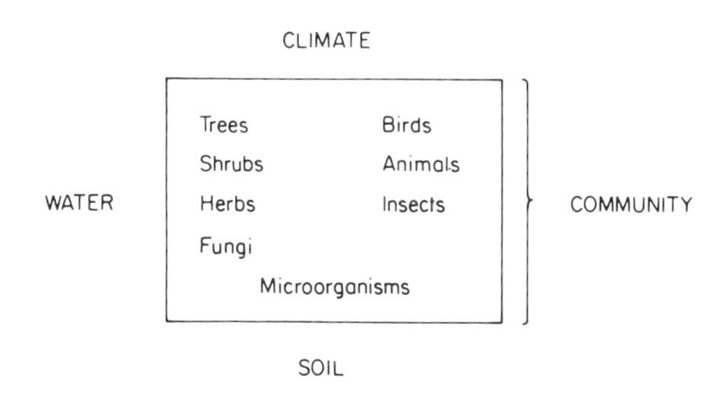

Fig. 1.1
The constituents of an ecosystem

Large, characteristic areas of vegetation, such as deserts, tropical rain forest, temperate grasslands, etc. are termed *biomes*. The tropical biomes are shown in fig. 1.3. Although dominated by one type of ecosystem, biomes can contain other types. For example, the tropical rain forest can include marshes, clearings, etc.

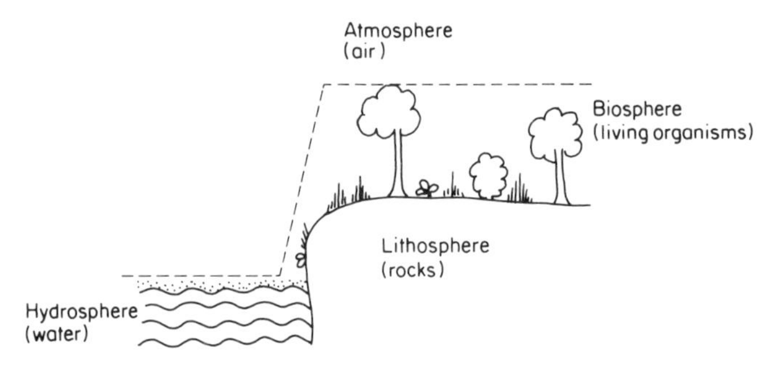

Fig. 1.2
The biosphere in relation to its environment

Within an ecosystem there are various *levels of organization* (fig. 1.4). A community can be broken down into *populations* of particular species and further into individual *organisms*. The study of an individual organism and its constituent parts is usually termed biology, while ecology studies populations and communities. Ecology and biology overlap, however, at the level of the individual organism, its relationship with its environment being termed *auto-ecology*.

It is often useful to talk about the total weight of living organisms in an ecosystem. This is termed the *biomass* and is defined as the weight of organisms per unit area.

A complete study of ecology covers not only such natural ecosystems as forests, deserts, etc., but man-made ecosystems, including cities, towns, dams and cultivated land. The majority of the ecosystems described in this book are natural, the influence of man on such ecosystems being covered in Chapter 9.

The components of an ecosystem can be divided into four main categories – *abiotic* (rocks, soil, water, climate); *autotrophs* (producers); *heterotrophs* (consumers), and *decomposers* (fungi, bacteria etc.). Autotroph means self-feeding and autotrophs are the plants and microorganisms which can convert inorganic substances into organic compounds utilizing the energy from the sun. Most autotrophs contain chlorophyll and convert carbon dioxide and water into sugars. Heterotrophs (other-feeders) have to obtain their food by eating other organisms. The *primary consumers* or herbivores eat plants, while the *secondary consumers* or carnivores eat herbivores. The decomposers obtain their food from dead material. A *food chain* can be constructed linking the various components of the living element of an ecosystem, as shown in fig. 1.5.

The food chain also illustrates the *energy flow* within an ecosystem, as energy from the sun is converted to chemical energy by the autotrophs and then passed along the chain to the decomposers. However, at every stage of the chain energy is lost, mainly as heat, and thus energy flow is a one-way process (fig. 1.5), unlike the cycling of nutrients.

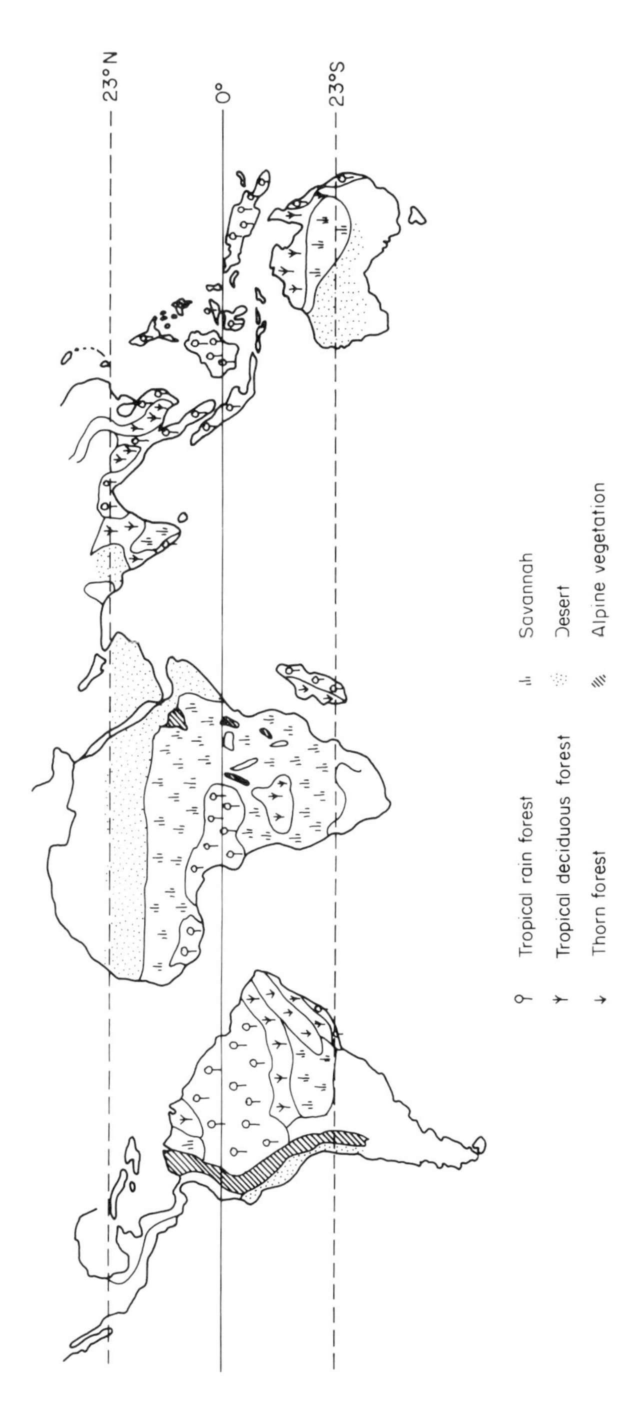

Fig. 1.3
Tropical biomes

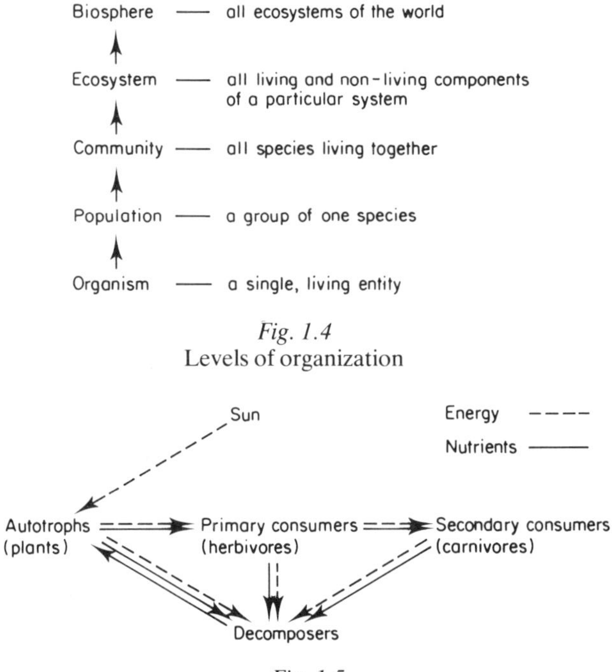

Fig. 1.4
Levels of organization

Fig. 1.5
The energy flow and cycling of nutrients within an ecosystem

Productivity

Only a small portion of the solar energy entering an ecosystem is actually captured by photosynthesizing organisms (autotrophs) and converted to chemical energy. The rate at which solar energy is converted to chemical energy is known as the *gross primary productivity*. As this energy is locked up in the bonds of organic compounds it can be equated with the production of dry matter by autotrophs.

Dry matter is broken down and energy released during the process of respiration, which takes place in all living organisms. Thus, though accumulating dry matter through photosynthesis, autotrophs also lose it through respiration. The difference between the gross primary productivity and the energy/dry matter lost through respiration of autotrophs is known as the *net primary productivity* of an ecosystem.

$$NPP = GPP - E_r$$

where NPP = net primary productivity
GPP = gross primary productivity
E_r = energy/dry matter lost through respiration of autotrophs

Net primary productivity is of great importance to an ecosystem as it represents the energy available to members further along the food chain. Ecosystems with small NPPs support fewer heterotrophs than those with high NPPs. In the tropics NPP can be roughly correlated with the amount of rain an area receives, although there are exceptions. Evergreen rain forests have much higher NPPs than savannas, as shown in Table 1.1.

Table 1
The net primary productivities of some tropical ecosystems

Tropical ecosystem	Average net primary productivity/g m^{-2} y^{-1}
Rain forest	2000–2200
Deciduous forest	1600
Savanna	700–900
Desert	100

The Biogeochemical Cycles

Two types of biogeochemical cycle occur – those such as the carbon and nitrogen cycles which contain a gas as the major constituent, and those, the sedimentary cycles, in which the major constituent is an insoluble compound. The gaseous cycles can be considered global, carbon dioxide and nitrogen being mobile and able to diffuse throughout the atmosphere. Thus, although there may be local pockets where nitrogen-containing compounds are over-abundant or in short supply, on a world-wide scale the concentration of nitrogen in the atmosphere and that locked up in organic and inorganic compounds remains fairly constant. The hydrological cycle, discussed in Chapter 3, also has a major gaseous constituent in the form of water vapour and can thus be considered global, despite the occurrence of oceans and deserts.

The sedimentary cycles, however, are localized and consist of many interwoven cycles, the reservoir of nutrient generally being locked up as an insoluble constituent of rocks and not mobile. The cycles of all mineral plant nutrients are sedimentary.

The carbon cycle

The carbon cycle (fig. 1.6) can be considered the most important of all the biogeochemical cycles, as ultimately all life on earth depends on its continued functioning. Primarily, the carbon (or carbon dioxide) cycle consists of two parts – the conversion of inorganic carbon dioxide to organic carbon compounds with the simultaneous storage of energy from sunlight, and the degradation of these compounds back to carbon dioxide with the release of this energy. The carbon cycle operates over land and water wherever green plants are to be found. Although small variations in carbon dioxide concentration are found locally, winds and the buffering effect of oceans which dissolve excess of the gas, ensure that globally carbon dioxide concentration in the atmosphere remains constant.

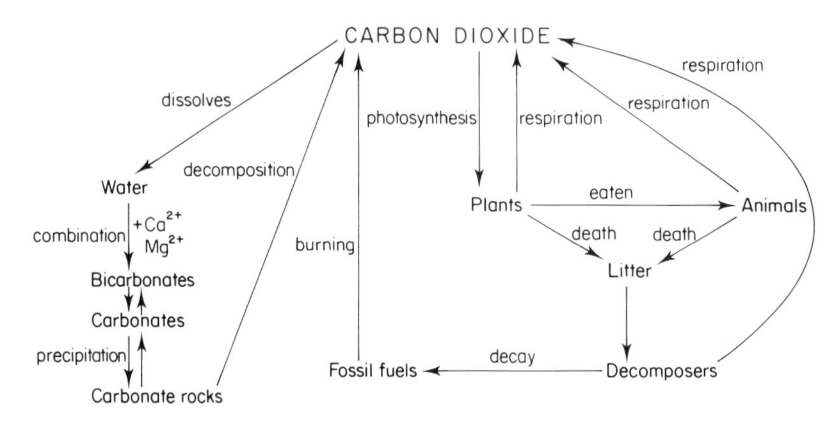

Fig. 1.6
The carbon cycle

The nitrogen cycle

Although some atmospheric nitrogen is fixed as oxides by electric discharges occurring during thunder storms, the conversion of the gas to inorganic nitrogen compounds is mainly carried out by nitrogen-fixing microorganisms (see Chapter 7). However, only a relatively small amount of nitrogen is locked up in organic and inorganic compounds at any one time and the cycle (fig. 1.7) depends on the efficiency of nitrifying and denitrifying microorganisms.

The phosphorus cycle

The phosphorus and other sedimentary cycles lack large, mobile reserves of elements and are therefore easily upset by natural processes or human interference. The phosphorus cycle (fig. 1.8) contains two main storage pools of the element – inorganic phosphorus compounds locked up in rocks and insoluble

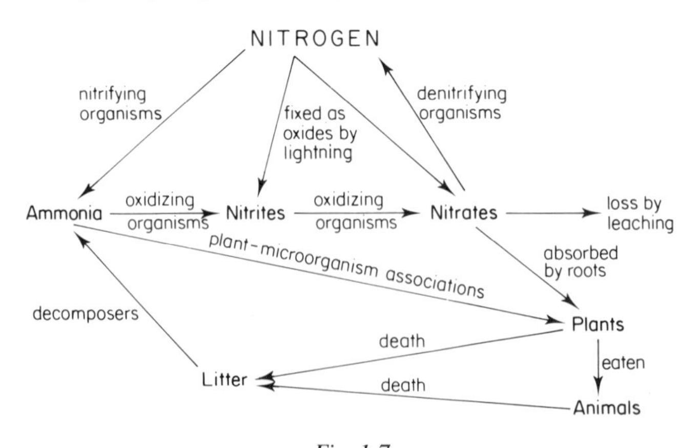

Fig. 1.7
The nitrogen cycle

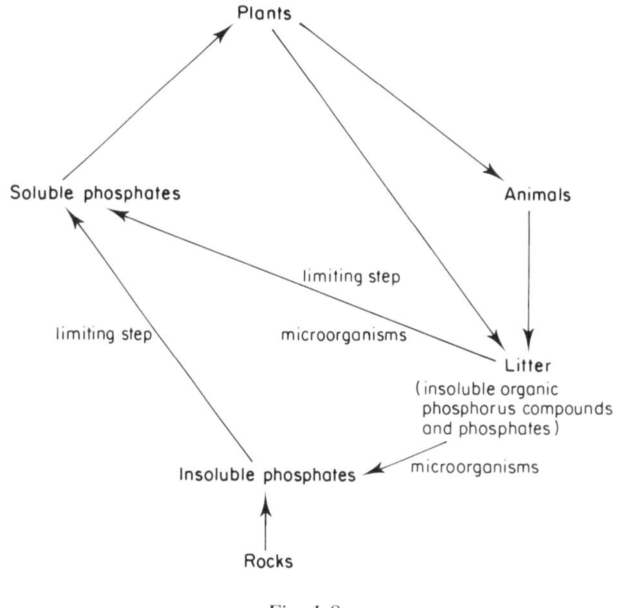

Fig. 1.8
The phosphorus cycle

organic compounds in litter. The solubility of inorganic phosphorus compounds and the degradation of litter by microorganisms are limiting steps in the cycle. Although little of the element is lost through leaching, erosion can cause serious depletion.

Succession

There are few ecosystems which are so stable that they show no appreciable change with time. Most are undergoing *succession* in which one set of species is replaced by another as the ecosystem develops. The colonization of bare rock, or new soil such as that resulting from a volcanic eruption or the deposit of windblown sand, are examples of *primary succession*. *Secondary succession* takes place on land which has supported vegetation in the past, such as abandoned fields, or the soil left behind after a forest fire or the death of a tree.

An ecosystem which has reached the stage where no more change takes place is said to be at its *climax*. Although no overall change of ecological significance takes place in an ecosystem at its climax, such a system is not static but is in dynamic equilibrium. Plants die and seeds germinate, old animals die and young animals are born, such processes taking place continuously in an ecosystem at its climax. A tropical rain forest is an example of primary succession at its climax, but within such forests secondary succession takes place whenever a tree dies and the area of soil once occupied by the tree is colonized by other species. The climax vegetation of secondary succession is often quite different from that of the previous climax.

Succession is the result of the modification of the environment by organisms occupying the habitat at any one time. Organisms which colonize an ecosystem during the first stages of its development are termed *colonizers* or *pioneer species*. Lichens colonizing bare rock, for example, are pioneer species which change the environment, making it suitable for the growth of other species. The fungal component of the lichen secretes chemical substances which dissolve some of the minerals contained in the rock. These dissolved nutrients are absorbed by the lichen, while the undissolved portion of rock forms pockets of soil in which small insects and microorganisms can live. Such organisms feed on dead organic matter and through their activities enrich the new soil, eventually converting it to a medium suitable for the growth of hardy, drought-tolerant plants (xerophytes). An increase in the depth of soil, and its enrichment through the death of pioneer species and the collection of wind-blown debris, eventually produces a medium which can hold sufficient water for the growth of less drought-tolerant species (mesophytes). Grasses are the most common of the pioneering mesophytes, and these, together with other species, form a *microclimate* at ground level, which is less extreme than that experienced by the xerophytes. The presence of such a microclimate enables shrub and eventually tree seedlings to develop. An example of the succession of an ecosystem is shown in fig. 1.9.

Each stage of succession changes the environment so that it becomes less suitable for the resident species and more suitable for invading species, until the climax is attained. Theoretically, the climax vegetation should remain in dynamic equilibrium forever, but in many instances it is upset by the activities of man or some natural catastrophe such as a hurricane, typhoon, or forest fire.

The environment of an ecosystem at its climax is quite different from that at the pioneer stage, as soil, temperature, humidity, light intensity, and windspeed are all affected by the vegetation within the ecosystem. The type of vegetation occurring at the climax depends on many factors, including climate, altitude, competition between species, soil type, and the activities of man. When the vegetation type depends only on the climate, the ecosystem is said to be at its *climatic climax*. Where factors such as soil determine the vegetation type, it is at an *edaphic climax*. Tropical rain forest is an example of a climatic climax, while marshland vegetation, which is the result of poor drainage, is an example of an edaphic climax.

At each stage of succession the biomass increases, as does the diversity of species, although the latter can decline in the final stages. The types of species found at any one time change rapidly during the first stages of succession but more slowly as the climax is approached.

Dominance

Generally, at each stage of succession there are some species which are more common than others and which are important in determining the nature of the ecosystem. Such species are termed *dominants*, and their removal radically

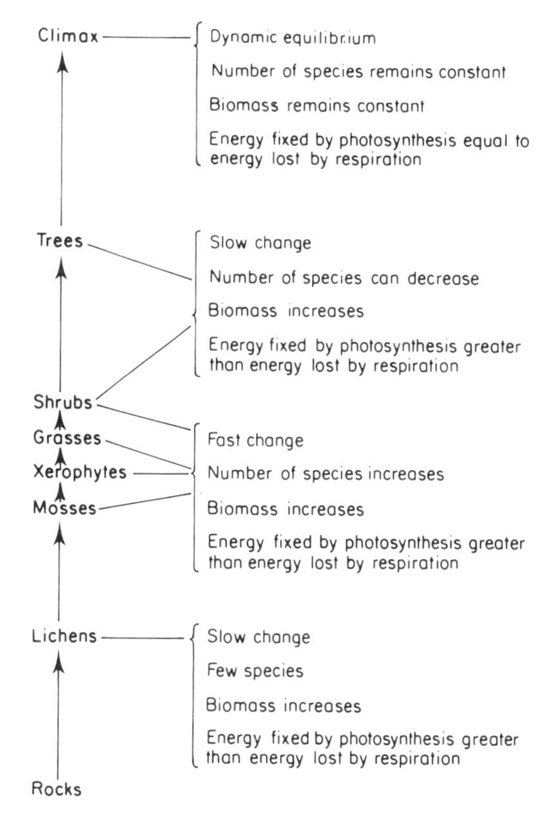

Fig. 1.9
Succession in an ecosystem

alters the ecosystem. The number of dominants in an ecosystem can vary widely. A typical tropical rain forest has a high diversity of species and can contain a dozen or more trees which are more plentiful than other species and can therefore be said to be dominant. Forest vegetation on an island, however, may be dominated by only one species, such as *Mora excelsa* in the West Indies.

Ecological Niche

Each organism within an ecosystem plays its own particular role within the overall scheme, this role being known as the *ecological niche* of the organism. The roles played by individual species are many and varied, but each is important to the overall success of the ecosystem. Green plants are producers and their main role is to convert energy from the sun into chemical energy which can be used to support other members of the community. Fungi, however, are the decomposers and their task is to release nutrients from dead organic matter for the use of green plants. Some insects are important as pollinators of flowers, ensuring future generations of plants, while others, such as termites, are decomposers which help to break down plant matter. Animals are also important as pollinators or seed dispersal agents. Today, the role of man is probably the most important of all, as his activities can determine the very nature of the ecosystem.

The kinds of organisms found in any ecosystem depend on geography and environment. Tree species found in the tropical rain forests of Asia are different from those found in similar forests in South America, but they occupy the same ecological niche. Thus ecologically equivalent species have evolved in different parts of the world with similar climates.

Species occupying different ecological niches in the same ecosystem live together quite happily, but those occupying the same niche are in competition with one another. Sometimes such competition leads to the extinction of one of the species from that particular ecosystem. By introducing or removing species man can influence the occupancy of an ecological niche.

The most successful species are those which can adapt to occupy different niches under varying sets of environmental conditions. Many plants have spread throughout the world, as they have been able to adapt to local habitats. Other species, unable to adapt, are found only in one specific type of environment. Many tropical plants, for example, are unable to adapt to low temperatures and are thus unable to grow in temperate climates. Often a species which cannot compete successfully for a niche under one set of environmental conditions adapts to a less favourable environment and thus occupies a different niche. Many plants which flourish on soils with a high calcium content (calcicoles) are unable to compete with other species on more neutral soils.

Some species occupy a very precise ecological niche, such as those insects associated with one particular plant species, while others have wider niches. The specialists are the most successful at exploiting their resources, but more vulnerable to change. In the above example, should the plant become extinct then so would the insect. The wider the niche the more likely an organism is to survive a drastic change in its environment.

Suggestions for Further Reading

Etherington, J. R. (1978). *Plant Physiological Ecology*. Arnold.
Grime, J. P. (1979). *Plant Strategies and Vegetation Processes*. Wiley.
Kormandy, E. J. (1976). *Concepts of Ecology* (second edition). Prentice-Hall.
Odum, E. P. (1975). *Ecology* (second edition). Holt, Rinehart and Winston.
Tivy, J. (1971). *Biogeography*. Oliver and Boyd.

Chapter 2

Plants and Soil

Introduction

Soil can be defined as the weathered layer of the earth's crust to which has been added the decomposition products of living and dead organisms. Mixed with the soil are air, water, and a multitude of living organisms, including algae, bacteria, fungi, the roots of plants, and many soil animals and insects. Soils are classified according to their *profiles*, which are obtained by cutting vertical sections through the soil from topsoil to underlying parent material. Such a section can be divided into *horizons*, the most important being the topsoil or A horizon, the subsoil or B horizon, and the weathered substratum or C horizon. A layer of litter usually covers the topsoil. The main features of a soil profile are shown in fig. 2.1. Often the A and B horizons are subdivided, while the litter layer on many tropical soils is thin, as high temperatures and humidity ensure that decomposition of this layer is rapid.

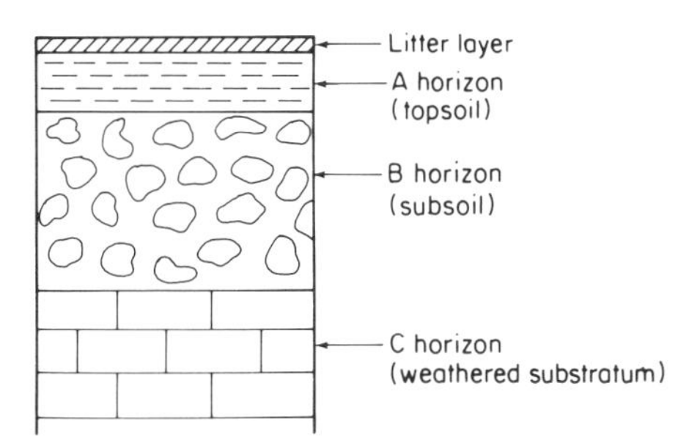

Fig. 2.1
The main features of a soil profile

The importance of soil to plants cannot be overstressed. Man may have devised the soilless method of cultivation known as hydroponics, but few plants grow naturally without soil. Parasites obtain their requirements from other plants and some lichens can grow directly on bare rock, but most plants need soil to grow to maturity. Seeds falling on bare rock or other soilless environments may germinate if water is present, but the young plants soon die. Soil provides nutrients, which are essential to growth, water, one of the basic requirements for photosynthesis, and air, which is needed by the roots for respiration. Soil also gives plants a medium in which roots can spread, thus anchoring the plant and increasing the available food supply. Plants growing in soils which hinder root penetration are poorly developed compared with those growing in soils in which the roots can easily spread. Because of the large surface area of the roots there is much contact between plant and soil. Not only does the soil affect the plant but the plant influences the soil, both physically and chemically. Penetration of the roots helps to break up large particles of soil, while roots secrete carbon dioxide and other substances which attack and break down minerals.

Because soil is such an important part of their environment plants have been classified according to the type of soil in which they naturally grow as shown in Table 2.1.

Table 2.1
Classification of plants according to soil type

Plants	Soil type
Oxylophytes	Acid soils
Calciphytes	Alkaline soils
Halophytes	Saline soils
Psammophytes	Sandy soils
Chasmophytes	Rock crevices
Lithophytes	Rock surfaces

A theoretical succession of plant types colonizing bare rock can be devised as shown in fig. 2.2, assuming that production of an acid soil is the final stage of succession.

Soils vary in the amounts of nutrients they contain and also in their ability to retain air and water. Despite the luxuriant vegetation of the tropical rain

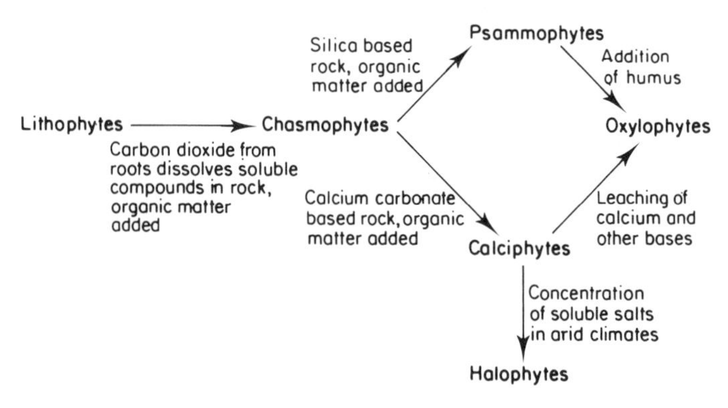

Fig. 2.2
Theoretical succession of plants colonizing bare rock

forests the soils usually contain little nutrients. The highly-weathered nature of such soils means that most of the original nutrients not absorbed by plants have been leached out. The continuous, rapid uptake and storage of nutrients by plants ensures that those released by decomposition of litter only remain in the soil for a short time. Streams draining areas of such highly-weathered soils have the conductivity of distilled water, although they may be coloured brown by humus washed out from the soil. In the tropical evergreen rain forests a continuous cycle of leaf fall, decay, and absorption of nutrients occurs, as shown in fig. 2.3.

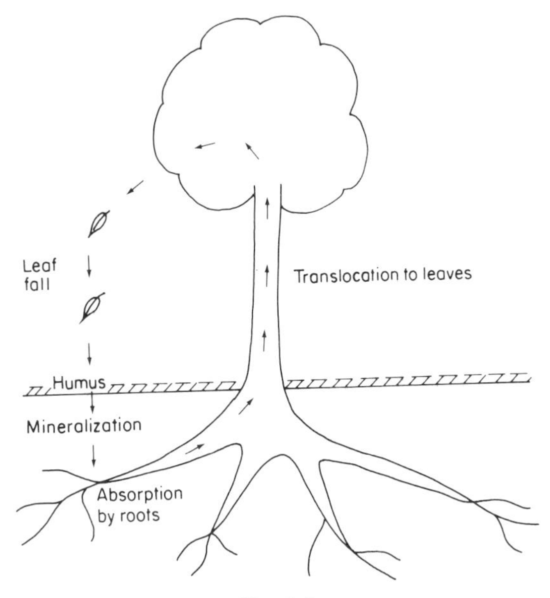

Fig. 2.3
The continuous cyclization of nutrients in the tropical rain forest

The clearing of tropical rain forest leads to intense leaching (Plate 2.1), so that unless artificial fertilizers are applied the soil can only support cultivation for a few years. If left fallow, a secondary forest develops which is less luxuriant and contains smaller trees than the primary forest. Undergrowth is denser, consisting of shrubs and many tangled creepers. In the past, periodic clearing of secondary forest for cultivation did little harm, as the forest was left undisturbed for many years. In some overpopulated areas of the tropics today, however, fallow times have been drastically reduced. Such practices eventually result in a soil which is too impoverished to support more than a few hardy grasses. These effects of overcultivation are summarized in fig. 2.4.

The Formation of Soil

The type of soil produced in any situation depends on at least seven factors – parent material, relief, climate, vegetation, soil organisms, time, and the

Plate 2.1
Erosion of soil after the removal of tropical rain forest. (Reproduced from Plants and Environment – A textbook of Autecology by R. F. Daubenmire, 3rd edition, 1974, by permission of John Wiley & Sons Inc.)

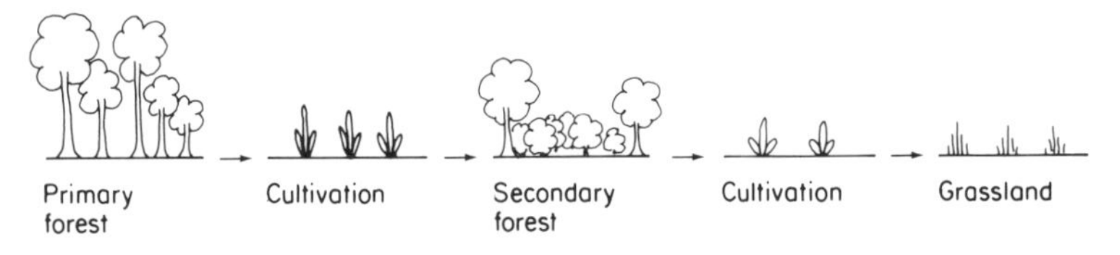

Fig. 2.4
The effects of shifting cultivation on tropical vegetation

activities of man. Although there have been attempts to separate these factors they are in fact interrelated, as shown in fig. 2.5. Most soils are in a stage of slow evolution. Some highly weathered soils of tropical rain forests undisturbed by man, however, have remained so little altered over such a considerable time that they can be considered to have reached a steady state and to be in dynamic equilibrium with their surroundings.

Physical and chemical weathering of rocks produces the parent material of soils, which is also known as the regolith. Physical weathering is of less importance in the tropics than in colder climates, where extremes of temperature help to break up the rock. However, some tropical plants, such as *Ficus umbellata* have roots which can penetrate small cracks in rocks, widening these

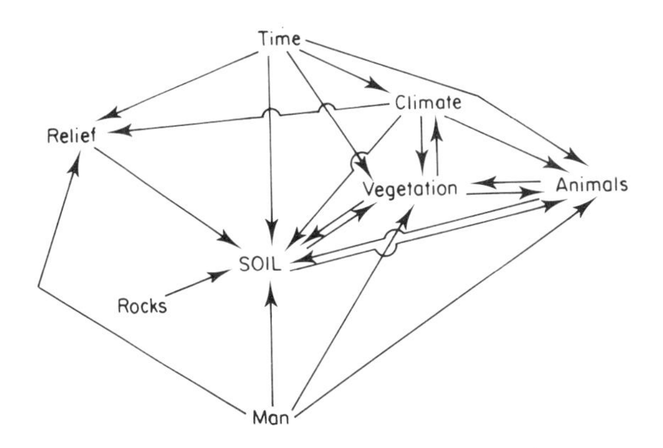

Fig. 2.5
Factors determining soil types

and eventually causing mechanical disintegration of the rock. The abrasive action of particles carried by water and wind also helps to erode large pieces of rock.

Chemical weathering is of great importance in the tropics and the regolith can be very deep, reaching 50 m or more. Chemical weathering is due to the processes of hydrolysis, oxidation, hydration, and carbonation. Much chemical weathering is a result of the action of acidic substances dissolved in the soil water. Atmospheric carbon dioxide dissolved in rain water, or that from the respiration of roots dissolved in soil water, is a weak acid, while organic acids are produced during the decomposition of litter. These acids attack rocks dissolving the bases and causing disintegration of the rock. The roots of the fungal symbiont of lichens excrete acidic substances which act in a similar manner, releasing nutrients for the plant and causing disintegration of the rock. On the death of the lichen the products of its decay enrich this new soil so that hardy xerophytes can grow. Eventually less hardy plants become established and the first colonizers die out. Chemical weathering converts primary minerals to secondary minerals, such as clays and solutes. The solutes may be removed entirely due to leaching by rainwater, as has happened in the tropical rain forests, or they may be washed downwards as in some tropical deciduous forests where rainfall is less.

Clays, the final products of physical and chemical weathering of the regolith, consist of particles of less than 0.002 mm diameter, larger particles being known as silt and sand. Silt particles have diameters of 0.5–0.002 mm, while those of sand have diameters of 2–0.5 mm. Soils with a large percentage of sand are well drained, but quickly suffer from drought (Plate 2.2). Those high in clay can be poorly drained but suffer less from drought. Loams, the best types of soil for most cultivation, are those in which sand, silt, and clay are mixed to give a well-drained soil, but one which retains sufficient moisture for plant needs.

Many tropical soils contain a high proportion of clay. Clays are formed both by weathering of the regolith and through combination of precipitated silicon

Plate 2.2
Sand dunes suffer from drought. (Reproduced from Plants and Environment – A textbook of Autecology by R. F. Daubenmire, 3rd edition, 1974, by permission of John Wiley & Sons Inc.)

and aluminium salts in the soil solution. Clay particles of less than 0.001 mm in diameter form colloids which complex with humus to form *micelles*. These micelles have a large surface area and are negatively charged so that they act as anions, attracting cations which are adsorbed onto their surfaces (fig. 2.6). These cations can be exchanged for other cations, the total number which can be so exchanged being known as the *cation exchange capacity* of the soil. The

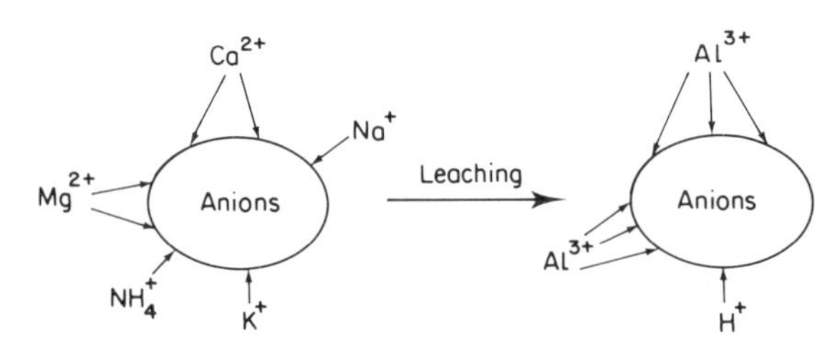

Fig. 2.6
Cation exchange on clay micelles

nutrient cations – calcium, potassium, ammonium, magnesium, and sodium are basic and can be easily displaced by hydrogen or aluminium cations, which are acidic. In highly-weathered tropical soils most of the nutrient cations have been replaced by aluminium or, to a lesser extent, hydrogen ions. Thus the concentration of basic cations is low and the soil acid.

The process whereby the nutrient ions are replaced and then washed out of the soil with the ground water is known as *leaching*. Under the hot, humid conditions of the tropics silica is also leached from some soils, but as leaching cannot take place under highly arid conditions many desert soils are saturated with sodium ions.

In the tropics two types of clay predominate: *kaolinite* and *montmorillonite*. Kaolinite has a 1:1 lattice structure in which alternating layers of silica and alumina are held together by hydrogen bonds (fig. 2.7). This gives a rigid structure which cannot absorb water molecules. Montmorillonite, however, has a 2:1 lattice structure in which an alumina sheet is held between two silica sheets, the bonds between the layers being weak, as shown in fig. 2.7. Such a structure is less rigid than that of kaolinite, and montmorillonite can absorb water molecules. Thus when wet, montmorillonite absorbs water and expands, while on drying it cracks, producing deep fissures as shown in fig. 2.9. Montmorillonite has a greater cation exchange capacity than kaolinite, as aluminium ions in the lattice may be replaced by magnesium, while cations can be held in the spaces between the layers. In kaolinite aluminium cannot be replaced by magnesium.

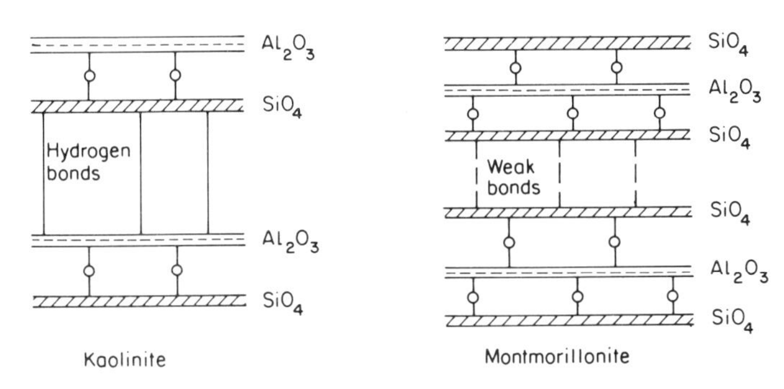

Fig. 2.7
The lattice structures of kaolinite and montmorillonite

The parent material of a soil need not be the bedrock beneath the soil. Soils which are derived from the bedrock beneath are known as *residual soils*, while those in which the parent material was derived from rocks elsewhere are known as *transported soils*. The main types of transported soils are *colluvial* (moved by gravity), *alluvial* (moved by water), and *eolian* (moved by wind). Examples of colluvial and alluvial soils as well as residual soils are shown by the *catenas* of the tropics. A catena (meaning chain) is the name given to an area of land in which the landscape changes regularly from hilltop to valley bottom. Many different types of catena exist; in that shown in fig. 2.8 the residual soil occurs on the hilltop and is thin due to erosion. Thus it can only support a sparse vegetation. Because this soil is well drained the iron left behind when the silica was leached out is in the form of ferric oxide, so that this hilltop soil is red. The soil on the hillside is mainly colluvial, the parent material being transported by gravity from the hilltop. This soil is both deeper and contains more water than the hilltop soil. The ferric oxide is in a hydrated form which gives the soil a

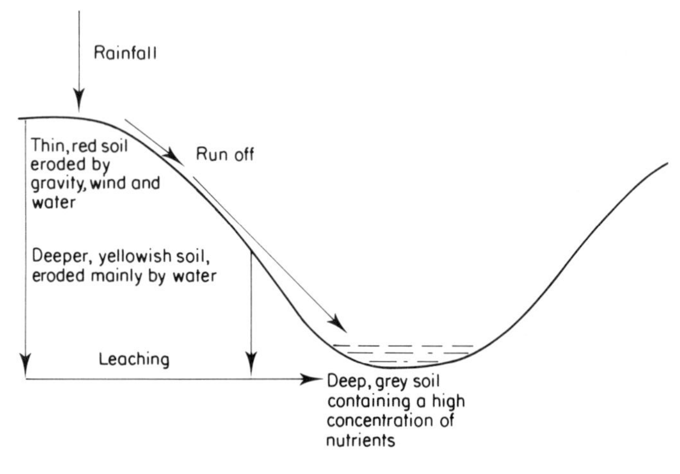

Fig. 2.8
An example of a catena

yellower colour. The parent material of the valley is alluvial, being washed down the hillside or the result of flooding by a river. It is badly drained and often waterlogged, so that the soil contains little air. Under these conditions iron compounds are in a reduced state and the soil has a grey colour. Because nutrients leached from the hills collect in the valleys the grey soils can be highly fertile under proper management. The three types of soil support different types of vegetation, the pattern being repeated over and over again throughout vast tracts of land.

Soil Structure

The structure of soil is important to plants as it can be responsible for the amount of air and water available and the ease with which roots can spread so as to tap fresh supplies of nutrients and water.

Most soils contain aggregates of particles which are held together by a coating of colloids derived from clay and humus. In the tropics such aggregates can be cemented together by iron or aluminium oxide films. These aggregates or *peds* have a high water and nutrient holding capacity, while the spaces around them ensure a good supply of air to the roots, which have little difficulty in penetrating these pores. Thus a well aggregated soil is the best type for plant growth.

Soils containing over 70 per cent sand are coarse textured, while those containing at least 60 per cent clay are fine textured. Rain can penetrate coarse-textured soils easily but much is lost as run off on fine-textured soils. The movement of water within the soil is also influenced by texture, fine-textured soils with their small interstitial spaces resisting infiltration. However, in coarse-textured soils, water penetrates downwards too quickly for shallow-rooted plants to survive unless rainfall is frequent or the water table high. Thus plants growing on soils with a high proportion of sand are usually deep rooted. The influence of soil structure on the availability of water and air to plants is further discussed in Chapters 3 and 5.

The Soil Water Table

Water held in the soil is very important to plants. Rain falling on the surface penetrates the soil until it reaches the *water table*, or layer of soil which is completely saturated with water. In well drained soils the water table usually lies just above the bedrock, but in badly-drained soils it can be much higher. The position of the water table also varies according to season and the amount of water entering the soil. In the tropics the water table is at its highest level at the end of the rains and at its lowest level at the beginning of the following rainy season. Soils containing layers of clay or hardened zones of iron and aluminium oxides beneath the surface may have hanging or perched water tables, as water is unable to penetrate such layers.

Tropical Soils

Many soil classification schemes have been proposed, but their discussion lies outside the scope of this book. In this section only the broadest distinction is drawn between soil types found in tropical areas. The zonal soils, such as the latosols and vertisols, depend primarily on climate, while the intrazonal and azonal soils reflect the dominance of a local factor, such as parent material or drainage conditions.

Latosols

The term *latosol* is used to describe a wide range of soils occurring in the humid tropics in which most of the free bases and silica have been removed by leaching. Such soils contain high concentrations of iron and aluminium oxides, free quartz grains, and kaolinite clays. They are red or pink and have a low organic content. The term laterite has also been used to describe such soils, but is now restricted to the hard concretions of iron and aluminium oxides described below. Latosols are also known as oxisols, ferralsols, or ferrallitic soils.

Latosols are very old and deeply weathered, the regolith being 50 m or more deep. They are present at elevations of less than 2000 m and are the predominant soils of the tropical evergreen rain forests. Only traces of primary, weatherable material remain, while leaching has removed most nutrient ions, so that the base saturation of these soils is low and the soil acid. Leaching of silica has left aluminium and ferric oxides, the latter giving such soils their characteristic red colour. Kaolinite is the type of clay found in latosols, aggregates being cemented together by ferric oxide.

Latosols support both tropical rain forest and savanna vegetation, depending on the rainfall pattern of the area. In the latter case the soils are less leached and not so acid.

The iron and aluminium oxides of latosols can become concentrated into a specific layer below the surface as a result of water movement. Removal of the vegetation cover results in erosion which brings this layer to the surface, when the action of air and sun produces an irreversible hardening, forming an iron pan or laterite. The term laterite is derived from the Latin *later*, meaning brick, as soft laterite can be cut into bricks which subsequently harden. In time, laterite becomes weathered to form a thin soil which can support a sparse vegetation. Theoretically this soil should eventually evolve to the stage when it can again support a forest vegetation. Some tropical crops, such as rubber (*Hevea brasiliensis*) can be grown on laterite, although smaller trees result.

Vertisols

Vertisols are formed in areas of the tropics with a pronounced dry season. They are brown or black in colour and are thus also known as the tropical black earths. These soils are formed from limestone, marl, or ferromagnesian rocks, and the predominant clay is montmorillonite. In the dry season vertisols shrink and crack producing fissures. Top soil falls into these cracks thus mixing the horizons. During the next rainy season the montmorillonite absorbs water and expands, closing the fissures. The pressure of the extra soil causes the formation of humps or microknolls on the surface, while microbasins appear where the cracks occurred (fig. 2.9).

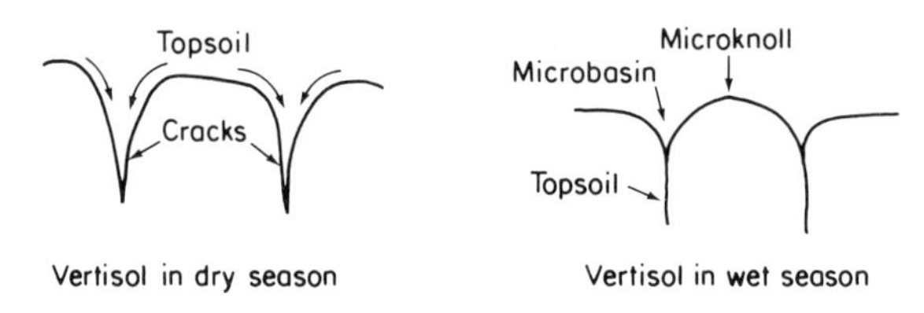

Fig. 2.9
Vertisols in wet and dry seasons

Podsols

Podsols are more characteristic of temperate climates than of the tropics, but such soils can occur where the parent material is very sandy. Podsols are formed when iron and aluminium are leached out leaving quartz sand, which forms a bleached horizon below the topsoil in the A horizon. Because they are highly leached and contain little calcium, podsols are infertile, their vegetation covering varying according to climate, situation, etc. In the tropics podsols are mostly found in coastal sand areas.

Hydromorphic soils

Hydromorphic or *gley* soils are formed under conditions of poor drainage such that water fills the pores around the soil aggregates driving out air. Under such anaerobic conditions ferric ions are reduced to the ferrous state, and as ferrous ions are soluble they are gradually leached out, leaving a grey coloured soil. During periods of drought such soils can become aerated when oxidation of ferrous ions to the ferric state occurs. The resulting ferric oxide produces red mottling in the B horizon.

Gley soils contain more nutrients than most other tropical soils and have therefore been much interfered with by man. Paddy fields for the cultivation of rice were derived from gley soils. However, badly constructed drainage schemes in the tropics have led to the production of soils containing a high concentration of sulphuric acid which are useless for cultivation. The sulphuric acid results from the oxidation of ferrous sulphide which is present in waterlogged soils.

Calcimorphic soils

Calcimorphic soils are formed over limestone. They are less leached, more alkaline, higher in bases, and more stable than soils of surrounding areas.

Halomorphic soils

During the weathering process rocks release large quantities of soluble salts, which in areas of high rainfall are leached out and carried away by streams. In areas of low rainfall, however, these salts are not washed away completely, although they may be leached downwards. In soils with a high water table the opposite may take place, the salts being brought to the surface as the soil solution rises by capillary action. Evaporation causes a layer of salts to be deposited on the surface of the ground, resulting in the formation of salt lakes (Plate 2.3).

Periods of high rainfall can result in salts being washed down from hillsides to collect in valleys, especially if the latter are badly drained. Similarly areas of high surface salinity can occur in depressions on arid tropical plains. Such salt licks are very important sources of minerals to the animals which inhabit such areas.

Azonal soils

The *azonal* soils can be found in any climatic region and are characterized by lack of a B horizon, the A horizon being thin and only differentiated from the C horizon by the presence of humus. Azonal soils include lithosols, regosols, alluvial soils or fluvisols, and organic soils. Lithosols occur on mountain tops, where the parent material is resistant to weathering. They are thin and easily eroded. Regosols are young soils formed from volcanic ash or sand. They are very mobile and easily eroded. Alluvial soils are deposited by water during flooding, while organic soils, such as peat, are rare in the tropics due to the rapid decomposition of organic matter.

Plate 2.3
Salt lake. (Reproduced by permission of A.A.A. photo, Paris.)

The Organic Constituents of Soils

The organic matter of soils is derived from both plants and animals. Material derived from dead roots and soil organisms is distributed throughout the soil, but the bulk of organic matter in a forest soil comes from the dead leaves and other plant parts which fall onto the surface and form the litter layer. Such litter is attacked by organisms which gradually break down the tissues to form *humus*. In the tropics the process of decay is rapid, so that the depth of the litter layer depends primarily on the rate at which dead organic material is deposited on the soil surface.

Humus is a sticky, dark brown, amorphous mixture of substances which has acid and colloidal properties. It is an important constituent of soils as it helps to bind the soil particles, increases water adsorption, and complexes with clay to form micelles, as described above. By binding soil particles into crumbs, humus increases the circulation of air and water. Humus is not the end product of decay, however, as it is slowly broken down to release nutrients, carbon dioxide, and water, a process known as *mineralization*. Under humid tropical conditions mineralization takes place more quickly than in colder or drier climates. The amount of humus in a soil depends both on the rate of mineralization and on that of litter deposition and breakdown. The humus content of savanna soils is generally low and is mainly derived from dead grass rootlets, the leaves mostly being either eaten by animals or destroyed by fire.

Soil organisms

The presence of organisms in the soil is very important to plants, and in general soils containing abundant organisms are also the most fertile. Organisms found in the soil include bacteria, fungi, algae, plant roots and other underground organs, protozoa, nemetodes, mites, worms, insects, and burrowing animals. One gram of soil may contain many millions of bacteria and up to a million protozoa, which prey on bacteria. Some protozoa live in the alimentary systems of termites and are responsible for the digestion of cellulose, an important stage in the breakdown of litter. The activity of termites is mainly responsible for the rapid breakdown of the litter layer of tropical soils.

Algae require sunlight for photosynthesis and therefore live on, or just beneath, the surface. The majority of microorganisms are held in the colloidal films coating soil particles, and bacteria, blue–green algae, and fungi all help to bind soil particles into aggregates. Soil fauna, such as termites and worms, help to improve soil structure and aeration by their movements.

Each stage in the breakdown of organic matter to form first humus and then mineral substances is carried out by a particular set of organisms, the overall process being extremely complicated. For example, proteins are first decomposed to amino acids and then to ammonium salts, the latter being oxidized first to nitrites and then to nitrates. Each stage of the process is carried out by a different organism.

Soil fauna, such as termites, worms, millipedes, etc., ingest organic matter and excrete it in a partly decomposed form, which is then acted upon by the various soil microorganisms. These microorganisms operate most effectively at temperatures around 25°C, the average found in the tropics. Thus the processes of decay are much more rapid than in temperate regions. Large concentrations of carbon dioxide or organic acids, such as may be found in poorly aerated soils can bring the processes of decay to a complete halt, as the microorganisms are unable to function under such conditions.

Some soil organisms, such as *Beijerinckia* in the tropics, can convert atmospheric nitrogen gas into nitrogen-containing salts, a process known as *nitrogen fixation*. Nitrogen fixed by blue–green algae living in the paddy fields is important to the cultivation of rice, while many leguminous plants can grow in soil with a low nitrogen content. These plants obtain their nitrogen compounds from the products of the nitrogen-fixing activities of the *Rhizobium* bacteria in their root nodules. Nitrogen fixation is discussed in more detail in Chapter 7.

However, the activities of soil organisms are not all beneficial to plants. A large number of bacteria, fungi, and insects prey on plants causing their eventual death, while under anaerobic conditions oxidation of humus to carbon dioxide and water does not take place and substances toxic to higher plants, such as acids, aldehydes, and phenols, are produced. Microorganisms also compete with plants for available nitrogen, which in a soil containing much decaying matter can result in nitrogen deficiency symptoms in higher plants. Although the movements of soil fauna are in general beneficial, they can result in a too rapid penetration of water, such that plants growing in the soil suffer from drought.

The living roots of plants affect the soil they are in by excreting carbon dioxide and by absorbing oxygen, water and nutrients. Both the excretion of carbon dioxide and the absorption of water affect the processes of leaching.

Roots

The roots produced by tropical trees and shrubs are admirably adapted to the environment experienced by the plants. Rain forest evergreen trees generally have extensive surface roots which do not penetrate the soil very deeply. Such roots are able to make the best use of nutrients released during the decay of litter and those washed into the soil by rain. Many such roots do not have root hairs but absorb nutrients through association with fungi (see Chapter 7). Lack of water is not a problem of rain forest trees and thus their roots do not need to penetrate the soil deeply in search of moisture. It is characteristic of many rain forest trees that they develop part of their root system above ground in the form of buttresses, which help to anchor the tree. Trees with large buttresses include *Mora excelsa* and *Piptadeniastrum africanum*.

In areas where the surface layers of soil suffer periodically from drought, plants with deeper penetrating root systems occur. These usually consist of both surface roots and a tap root which can reach moisture below surface level. In highly arid areas, shrubs with long tap roots but no surface roots are characteristic. These tap roots are able to penetrate the wet area (capillary fringe) which lies just above the water table.

The roots produced by trees growing in swamps are adapted to shifting soil and low oxygen conditions. The characteristic stilt roots of such plants help to anchor the trees, while the various types of pneumatophores, which extend above the water or mud levels, ensure that sufficient oxygen is available to the plant. Such roots are further discussed in Chapter 5.

Soil Nutrients

All plants require nine macronutrients and seven micronutrients for growth. The macronutrients are carbon, hydrogen, oxygen, nitrogen, phosphorus, potassium, sulphur, calcium, and magnesium, while the micronutrients are copper, zinc, boron, chlorine, molybdenum, manganese, and iron. Carbon, hydrogen, and oxygen are obtained from the air and water, but all other nutrients must be absorbed from the soil. The origin of the macronutrients, the ions absorbed by plants and experimental deficiency symptoms, are given in Table 2.2.

Table 2.2
The macronutrients absorbed by plants and experimental deficiency symptoms

Nutrient	Form in the soil	Ions absorbed	Deficiency symptoms
Nitrogen	Organic compounds	NO_3^-	Reduced growth, yellowing or reddening of leaves
Phosphorus	Organic compounds, inorganic phosphates	$H_2PO_4^-$, HPO_4^{2-}	Delayed flowering, reduced growth, darkening of leaves
Sulphur	Organic compounds, inorganic sulphates	SO_4^{2-}	Similar to nitrogen, chlorosis of young leaves
Potassium	Clay minerals	K^+	Interference with water balance, root rot
Magnesium	Magnesium salts	Mg^{2+}	Reduced growth, chlorosis of old leaves
Calcium	Calcium salts	Ca^{2+}	Interference with growth, leaf deformation

Of all the plant nutrients only nitrogen is not present in the rocks from which the parent materials of soils are derived. Some atmospheric nitrogen is fixed as inorganic compounds during thunder storms and is washed into the soil with the rain. However, most soil nitrogen orginates from the breakdown of organic matter and much is obtained from the dead bodies of nitrogen-fixing organisms (see Chapter 7), especially in newly developing soils. The fixation of atmospheric nitrogen by the blue–green algal symbiont of lichens is as important to soil development as the chemical weathering of rock by the fungal symbiont. A large number of organisms can convert organic nitrogen to ammonia, but in general this must be converted to nitrates by nitrifying bacteria before it can be absorbed by higher plants. The efficiency of nitrifying bacteria is impaired by high concentrations of carbon dioxide or organic acids, such as those produced by fungi decomposing organic matter. This would appear to be a regulating mechanism preventing the accumulation of high concentrations of soluble nitrates in the soil at any one time. Such an accumulation would be wasteful as nitrates are easily leached out of the soil and therefore lost to the ecosystem.

Most of the metallic elements in the soil needed by plants exist in three forms – sparingly soluble inorganic or organic compounds, cations absorbed onto the clay–humus complex, or as free cations in the soil solution. In areas of high rainfall, such as the humid tropics, ions in the soil solution are rapidly leached out. However, they are replaced by ions from the cation exchange complex at rates which enable plant roots to absorb them before leaching occurs (fig. 2.10a). Thus the exchange complex acts as a storehouse of plant nutrients and the total concentration of cations available to plants is the sum of those present in the soil solution and those held on the exchange complex.

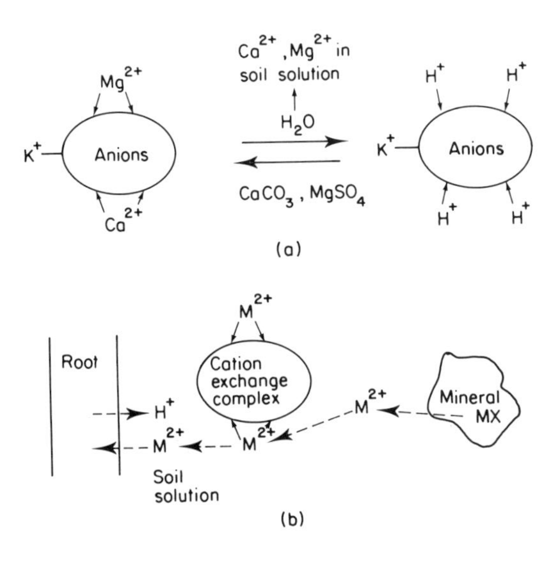

Fig. 2.10
(A) The exchange of cations on the surface of the clay–humus complex; (B) The absorption and replacement of cations

Release of cations from insoluble compounds is an even slower process, so that even under such adverse conditions as are found in the humid tropics, a regular supply of nutrients is still maintained (fig. 2.10b). Deeply rooted plants are able to absorb nutrients which have been leached downwards. These nutrients are eventually deposited on the surface of the soil when leaves are shed or the plant dies, thus increasing their availability to shallower-rooted plants (fig. 2.11). Rainwater also washes nutrients from the atmosphere and from leaves into the topsoil.

Some plants are able to absorb nutrients directly from the breakdown of organic matter. This is achieved through the association of their roots with mycorrhizal fungi.

Potassium is continuously recycled, as compounds of this element are very soluble and easily leached. They are washed out of living leaves by rain and enter the soil to be reabsorbed by the roots. Most soil potassium, however, is held as non-exchangeable cations in the lattices of clays; very little is held in organic compounds.

The concentration of calcium in soils varies widely and depends on the bedrock and parent material. Not only is calcium important as a plant nutrient

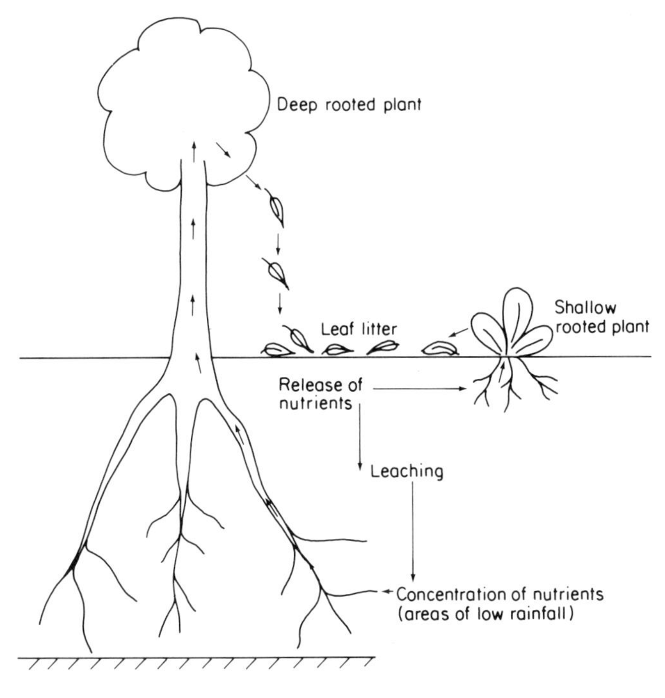

Fig. 2.11
Nutrient cycles

but it also influences the pH and aggregation of the soil. In the tropics the concentration of calcium in latosols is low, but some vertisols contain this element in high concentration.

The anionic phosphates and sulphates are derived from both the parent material of the soil and from organic matter. The availability of phosphorus depends on the pH and is discussed in the following section. Most tropical soils are low in phosphorus and even if added as fertilizer it soon becomes converted to forms unavailable to plants. Sulphate concentrations can be increased through the oxidation of hydrogen sulphide or sulphur dioxide. Hydrogen sulphide is formed naturally in anaerobic conditions such as waterlogged soils, while sulphur dioxide is one of the common pollutants.

The micronutrients all originate from parent material and deficiencies of these elements may occur. Some parent materials are deficient in copper and zinc, while light, sandy soils are often deficient in boron. In arid, saline soils, however, plants can suffer from toxicity symptoms due to too high a concentration of boron.

Sometimes the lack of a nutrient or the presence of an element at levels toxic to most plants is the deciding factor in determining the vegetation cover of a particular soil. Some wild species can become adapted to nutrient deficiencies quite quickly, but in general this is not true of cultivated plants. The natural slow growth of some plants may also be an adaptation to nutrient deficiency, especially nitrogen, sulphur, or phosphorus. Most newly formed soils are deficient in some essential element and pioneering plants are able to concentrate the little that is available. On their death they enrich the soil so that plants which cannot tolerate such a deficiency are able to grow. Compounds dissolved in rain water or contained in wind-blown debris also help to counteract deficiencies.

Some plants have become adapted to high levels of elements which are toxic to other plants or to animals. Such plants are usually termed *accumulators*. Many plants belonging to the Rubiaceae and Melastomataceae families accumulate large amounts of aluminium. Such plants are abundant on latosols, which contain much aluminium, but are rare on podsols, in which the content of this element is low. *Astragalus* and some other leguminous plants will accumulate selenium if grown in soils high in this element, while other plants are good indicators of metal ores. In Central Africa, for example, *Buchnera cupricola*, *Guttenbergia cupricola*, and *Becium homblei* are indicator species for copper ores as they tolerate a high concentration of the element. In Zambia the presence of *Mechovia grandiflora* is a good indication of the occurrence of manganese, while *Acrocephalus robertii* is an indicator for cobalt. The plants *Pearsonia metallifera* and *Convolvulus ocellatus* are indicator species for chromium ores in Zimbabwe.

Plant roots are able to absorb ions in a selective manner, often against concentration gradients. For example, potassium is usually present in the soil at much lower concentrations than sodium. However, plants absorb potassium ions to a much greater extent than sodium ions, the potassium concentration in plant sap being many hundred times that of the soil solution. Some nutrients appear to be absorbed directly by plants without first entering the soil solution. It is probable that the microorganisms of the rhizosphere help in this process.

The mechanisms of absorption are still open to question, but it has been shown that both diffusion and active pumping, similar to the pumps of animal cells, occurs. Active pumping requires energy which is obtained by the process of respiration and is one of the reasons why roots need a good supply of oxygen. Because of energy relations plants cannot absorb ions without an equal number of similar charge being released. Those released during the absorption of anions are usually carbonate or bicarbonate. Exchange processes, such as nitrate for bicarbonate, take place at the root surface, the nitrate being subsequently carried to the leaves by the transpiration system. During metabolic processes organic acids are formed which are transported to the roots in the phloem, where they are converted to bicarbonate, thus maintaining the cycle (fig. 2.12). In order to preserve neutrality the anions are transported in conjunction with potassium ions.

Mixtures of ions in the soil solution at the root surfaces can act synergistically or antagonistically. Calcium, for instance, stimulates the uptake of sodium. Microorganisms in the rhizosphere can also influence the uptake of nutrients, particularly that of phosphorus.

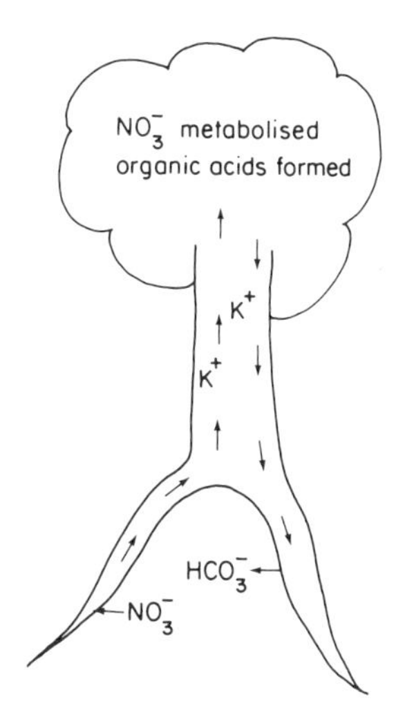

Fig. 2.12
The exchange of nitrate for bicarbonate at root surfaces

The Acidic and Basic Properties of Soils

The acidity or basicity of a soil depends on the hydrogen ion concentration in the soil solution. Hydrogen ion concentration is always very small, that of pure water being 10^{-7} mole per litre. Because the actual values of hydrogen ion concentration are cumbersome to use, a scale, known as the pH, was devised using only the indices without their negative signs. Thus water has a pH of 7. Acidic solutions contain more hydrogen ions than water, the hydrogen ion concentration of a typical organic acid being 10^{-5} mole per litre and its pH equal to 5. Alkaline solutions, however, contain fewer hydrogen ions than water, so they have pH values greater than 7. Thus an indication of the acidity or basicity of a soil is given by its pH value, those with a low pH being acid, while those with a high pH are alkaline. Most soils lie within a pH range of 3–8, those with a pH of 7 being termed neutral soils.

The extent to which the pH of a soil changes with the addition of acidic or basic substances depends on its constituents. Light, sandy soils are most affected, while soils with a high clay content show least change, due to the hydrogen ion absoring properties of the clay complexes. Addition of acidic substances to clay soils results in the adsorption of hydrogen ions, while addition of basic substances causes the release of hydrogen ions. In acid soils of pH 5 or less, aluminium ions behave in a similar way to hydrogen ions. This resistance to change in pH is known as *buffering* and clay soils are said to be well buffered.

Most plants suffer nutritional deficiencies when the pH of the soil lies outside the range 5–7, but even in highly acid or basic soils plants are found which have become adapted to such extreme conditions. In general, acid soils cause a wider range of stress symptoms to plants than alkaline soils. Acidic soils are generally deficient in calcium, magnesium, and potassium. The concentration of nitrogen may also be low, as bacteria involved in fixing this nutrient require calcium in order to function. Phosphorus is less available to plants in both acid and basic soils than in neutral soils. In acid soils insoluble ferric or aluminium phosphates are formed, while in alkaline soils available phosphorus is removed as insoluble calcium phosphate. Molybdenum becomes unavailable in acid soils, while the increased solubility of iron, aluminium, manganese, and some heavy metal compounds can lead to toxicity symptoms in plants.

Moderately alkaline soils cause less stress symptoms to plants as the presence of calcium has several beneficial effects. It increases the aggregation of soil particles and thus the aeration and water flow within the soil. Calcium is also an essential nutrient for the bacteria involved in nitrogen fixation and in converting ammonia to nitrates.

Plants which have become adapted to extremes of soil pH are known as *calcicoles* (calcium loving) and *calcifuges* (calcium hating). Calcicoles or calciphytes need large concentrations of calcium and cannot tolerate the concentrations of aluminium found in soils of low pH. Calcifuges or oxylophytes require only small amounts of calcium and in high concentration this element becomes a toxin. These plants can, however, tolerate high concentrations of aluminium.

Many soils experiencing average or low rainfall change in pH with depth, the topsoil being acid while lower horizons are alkaline. This is due to the leaching of calcium which is washed downwards, but rainfall is insufficient to wash this ion out of the soil altogether. The vegetation of such soils often contains both shallow-rooted calcifuges and deep-rooted calcicoles (fig. 2.13).

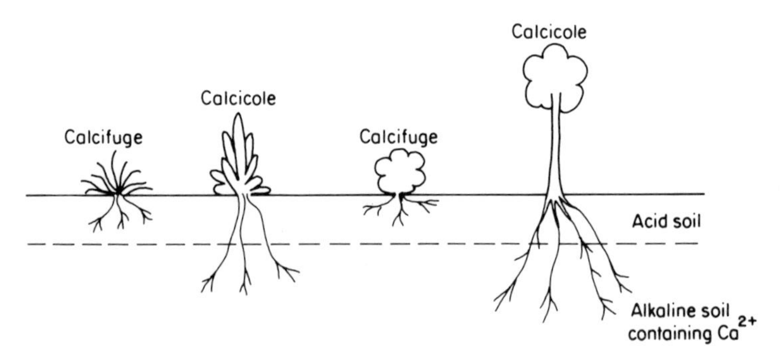

Fig. 2.13
The change of soil pH with depth in arid regions

Suggestions for Further Reading

Ahn, P. M. (1970). *West African Soils*. Oxford University Press.
Bowen, H. J. M. (1979). *Environmental Chemistry of the Elements*. Academic Press.
Bridges, E. M. (1978). *World Soils*. Cambridge University Press.

Etherington, J. R. (1975). *Environment and Plant Ecology*. Wiley.
Fitzpatrick, E. A. (1980). *Soils*. Longman.
Larcher, W. (1980). *Physiological Plant Ecology*. Springer-Verlag.
Longman, K. A. and Jenik, J. (1974). *Tropical Forest and its Environment*. Longman.
Richards, P. W. (1952). *Tropical Rain Forest*. Cambridge University Press.
Sutcliffe, J. F. and Baker, D. A. (1981). *Plants and Mineral Salts*. (second edition), Arnold.
Tivy, J. (1971). *Biogeography*. Oliver and Boyd.
Walter, H. (1971). *Ecology of Tropical and Sub-tropical Vegetation*. Oliver and Boyd.

Chapter 3

Plants and Water

Introduction

Water is of paramount importance to plants, as it is to all living organisms. Plant cells can contain as much as 90 per cent water, and few can survive drying to more than 40 per cent of their normal water content, while most protoplasm dies if the water content is reduced below 10 per cent. Water is also one of the raw materials of photosynthesis (see Chapter 4) and is the medium by which nutrients are absorbed. Water is a universal solvent and will dissolve all the chemical compounds required by plants, while its electrical properties allow these compounds to dissociate into ions, the form in which they are absorbed by the roots and transported within the plant. Water is also the medium in which the majority of chemical reactions take place within the plant.

The physical properties of water are important to plants in that water maintains the turgidity of cells which is necessary for their functioning, and especially for photosynthesis to take place. This solvent is also able to absorb heat in such a manner that the temperature does not rise sharply. Thus small temperature fluctuations in the external surroundings have little effect within the plant. This allows biochemical reactions to take place under relatively uniform conditions.

Most land plants obtain their water from the soil, although some absorption can take place through the leaves and other green parts. Some mosses, lichens, and algae, however, obtain their water directly from the air and are able to survive extreme desiccation.

Water within plants occurs in several forms. It is a chemically bound constituent of protoplasm, is stored in the vacuoles, and occurs as water of hydration. Interstitial water is the medium by which dissolved substances are transported.

Water of hydration consists of water molecules bound strongly to ions and organic molecules by electrostatic forces. Ions and polar groups, such as hydroxyl, amino, and carboxyl, attract water so strongly that a layer several molecules thick surrounds each ion or group (fig. 3.1). The thickness of this layer depends on the size of the ion and its charge. The water in these layers has a crystalline structure and is difficult to remove. Water of hydration is essential to life and accounts for 5–10 per cent of the total cell water.

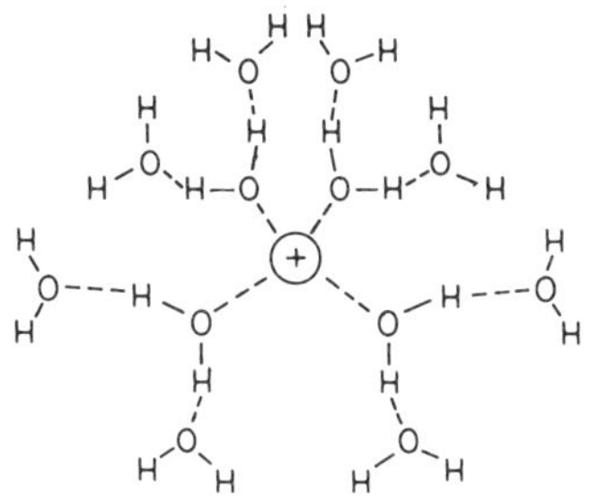

Fig. 3.1
An example of water of hydration

External water is important to plants both as visible and invisible vapour, as well as in the liquid form.

The Hydrological Cycle

Water enters the soil in the form of rain, dew or melting snow and ice and is removed from the soil by evaporation and the transpiration systems of plants. Thus a cyclic system is set up, as shown in fig. 3.2, which is known as the *water cycle* or *hydrological cycle*. Direct evaporation from soil with a covering of vegetation is small compared with the loss of water by transpiration and thus plants are important components of the hydrological cycle.

Precipitation

Water vapour can be precipitated in a number of ways, which include rain, snow, sleet, hail, dew, and mist droplets. The importance of dew and mist droplets will be discussed below, while snow and sleet are only experienced in the tropics at very high altitudes, where little or no vegetation occurs. Hail, which can cause considerable damage to plants, sometimes occurs during thunder storms in the tropics. Being of short duration, however, it contributes little to soil moisture. Rain is the most important of all the environmental factors affecting the type of vegetation to be found in almost any region of the tropics.

The annual rainfall varies over a very wide range from less than 100 mm in the deserts to over 10,000 mm in some tropical rain forests. Theoretically the seasonal distribution of rain should depend on the distance from the equator, as shown in fig. 3.3. When the sun is overhead convectional air currents result in the formation of cloud and the precipitation of rain. Thus regions on the equator should have no dry season. Due to the rotation of the earth those regions near the tropics of Cancer and Capricorn, at 23.5° north and south respectively, have the sun overhead only once per year and thus experience one long dry season and one wet season (Plate 3.1). Between these extremes the

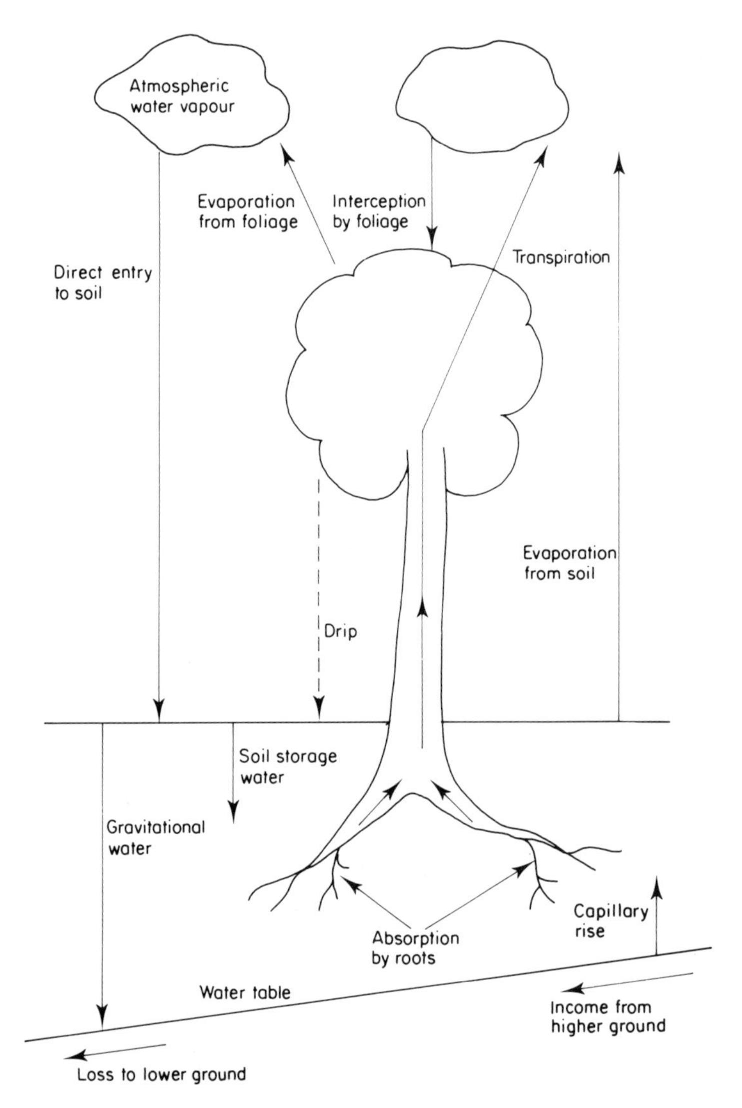

Fig. 3.2
The hydrological cycle

sun passes overhead twice a year, so that a gradual change from two short dry seasons to one short and one long dry season is found. However, these patterns of rainfall distribution can be considerably modified by other environmental factors, such as nearness to the sea, the topography of the terrain, or the occurrence of monsoons.

Most of the precipitation that falls on land results from water vapour derived from the surfaces of the oceans. Landward breezes forced abruptly upward by hills or mountains are rapidly cooled and thus lose a high proportion of their

Distance from equator		
23°	Sun overhead once a year	One long dry season
	Sun overhead twice a year	One short and one long dry season
	Sun overhead except for two short periods	Two short dry seasons
0°	Sun always overhead	No definite dry season

Fig. 3.3
The theoretical distribution of rain in the tropics

Plate 3.1
The alternate flooding and desiccation of this soil surface results in a habitat too severe for flowering plants. (Reproduced from Plants and Environment – A textbook of Autecology by R. F. Daubenmire, 3rd edition, 1974, by permission of John Wiley & Sons Inc.)

water vapour as rain. Descending air is warmed and therefore yields little or no rain on the leeward sides of hills or mountains. The abrupt change in vegetation that such directional differences can produce is well illustrated in the areas bordering the Rift Valley in East Africa. Lush vegetation is found on the windward slopes, known as the Highlands, which experience an annual rainfall of 1000 mm or more. The leeward slopes, however, can receive less than 500 mm of rain per year and support only a scrubby, desert-like vegetation dominated by the tree-like candelabra *Euphorbia* species (Plate 3.2).

Plate 3.2

A xeromorphic *Euphorbia* species. (Reproduced from Plants and Environment – A textbook of Autecology by R. F. Daubenmire, 3rd edition, 1974, by permission of John Wiley & Sons Inc.)

The topography of islands is particularly important in determining the amount of rain they receive. Islands of level terrain often experience low rainfall, but the windward sides of mountainous islands can be very wet. For example, the station Kauai on Hawaii has recorded an average of 11,450 mm of rain per year. It is thus possible, through the effects of topography, to have rain forest vegetation on the windward side of a mountain range in close proximity to desert vegetation on the leeward side, in areas where the wind direction shows little change throughout the year. Such contrasting vegetation types are found in the West Indies in Jamaica and Cuba, where the prevailing easterly trade winds are the cause of high rainfall on the windward slopes.

The seasonal distribution of rainfall influences the type of vegetation to be found in any area, rather than the actual amount of rain received during the year. Evergreen tropical rain forest is found near the equator where the climate is continuously wet, but on moving towards the tropics the vegetation changes to semi-evergreen forest in which the tallest trees are deciduous, losing their leaves during the short dry seasons. Lower canopies are able to remain evergreen due to the high relative humidity maintained within the forest. However, as the dry season becomes longer and the humidity reduced, only deciduous species of trees and shrubs can survive. In the deciduous forests the length of time trees are without leaves depends on the available water in the soil. Thus those growing in damp places, such as valleys, keep their leaves for a longer time than trees in more exposed positions.

As latitudes experiencing a single, long dry season are approached the vegetation changes to savanna, although strips of forest are still found along coasts and rivers. Grasses are well adapted to surviving a long dry season, growing vigorously while water is available, but the whole of the aerial system dies during the drought period. Only the roots and growing tip remain alive and dormant until the next rain falls, these being protected from desiccation by the layers of dead leaves. Woody plants cannot survive without some water, as although they lose their leaves the buds for next season's growth remain and continually lose water to the atmosphere. Thus a woody plant which is unable to absorb sufficient water from the soil soon dies, but grasses are able to withstand several months of drought. Evergreen trees growing in arid habitats have long tap roots which penetrate the capillary fringe. Thus, though the surface soil suffers from drought, water is always available to these plants.

Effective Precipitation

In general, savanna type vegetation is found in areas of the tropics receiving less than 2000 mm of rain per year, while desert conditions prevail where the annual rainfall is less than 300 mm. However, total rainfall is of less importance than the number of dry months and the amount of rain which is actually available to plants. The latter is known as the *effective precipitation*. The water contained in a shower of rain can be lost to plants in a number of ways. Some evaporates as it falls through the air, while more is intercepted by foliage and evaporated from leaf surfaces before it can be absorbed by the plant or drip to the ground. Even water which does reach the soil is not always available to plants. That from a light shower only penetrates a few centimetres and soon evaporates into the air again, while much water from a heavy storm is lost as runoff, especially on slopes. Water entering a highly porous soil percolates rapidly down to the water table and may soon be out of reach of plant roots. Thus effective precipitation P_E, the water which is available for absorption by plants, can be defined as:

$$P_E = R - E_a - E_s - E_p - O - G$$

where R = rainfall
E_a = evaporation while falling
E_s = evaporation from the surface of the soil
E_p = evaporation from the surfaces of plants
O = runoff
G = gravitational water

Humidity

The invisible water vapour in the air is termed the *humidity*, and is usually measured as *relative humidity*, which is the amount of water vapour in the air compared to the amount the air could hold at a specific temperature and pressure. Relative humidity RH is usually expressed as a percentage, thus:–

$$RH = \frac{H}{H_{tp}} \times 100$$

where H = absolute humidity (the amount of water actually in the air)
H_{tp} = maximum humidity at the same temperature and pressure

Except at high altitudes, the air pressure shows little variation. Thus H_{tp} depends on temperature and has a higher value in the tropics than in temperate regions.

Relative humidity normally undergoes a daily rhythm, being lower during the day than at night. However, in tropical rain forests the relative humidity remains near to 100 per cent throughout the day, whereas in deserts it can fall below 10 per cent. During the day the relative humidity usually decreases with distance above the ground, especially where there is a dense plant cover, as in forests.

Cloud, Mist, and Fog

The visible water vapour in the air occurs as cloud, mist, or fog. Cloud and fog or mist consist of water droplets or tiny ice crystals which result when the air cools to a temperature below its *dew-point*. The dew-point is the temperature at which the air becomes saturated with water. Clouds form when air rises to a cooler level, and they are usually separated from the ground. Fog or mist, however, are formed by cooling of the air at the earth's surface and are thus visible at ground level. The morning mists of the tropics are usually due to the radiation of heat at night, which cools the ground considerably when the air is still. Such mists rapidly disappear after sunrise.

The Importance of Atmospheric Vapours to Plants

Both visible and invisible water vapour affect plants in a number of ways. Solar radiation, the heat, light and other forms of energy which originate from the sun, is very important to plants (see Chapter 4). However, that reaching ground level depends to a large extent on the water vapour in the air. Visible water vapour intercepts much of this energy, reducing photosynthesis and transpiration. Thus plants growing in continual mist, as in the cloud forests discussed below, are often dwarfed and stunted compared to those experiencing only occasional fog.

When fog moves horizontally, or a cloud comes into contact with the earth's surface, minute drops of water are deposited on the leaves of plants. Such moisture may be absorbed directly, and on the rainless coast of Peru it serves as the sole source of water for plants. In less arid regions foliage may become so saturated that drops of water fall to the ground, increasing soil moisture. Such a situation exists in the cloud forests.

While most higher plants obtain their water requirements from precipitated moisture, some mosses and lichens are able to extract water vapour from air with a high humidity. In general, the abundance of such plants, especially those that grow on bare rock, depends on the humidity of the climate. Certain tropical orchids and members of the Bromeliaceae family growing as epiphytes on the branches of trees (see Chapter 6) can absorb water directly from the air when the relative humidity is above 85 per cent. Some desert plants also obtain their water supply directly from the air.

Whenever loss of heat by radiation cools a surface below the dew-point water vapour will condense on this surface, forming a film of dew. This moisture can be absorbed through the cuticle of normal epidermal cells or through the specialized organs of some plants. For example, the scales at the base of the leaves of epiphytic bromeliads absorb water collecting in the funnelled leaf bases, while epiphytic orchids absorb moisture through their aerial roots.

Coastal and riverine environments experience high humidities compared with surrounding areas, even during periods of drought. Such regions support plants which otherwise would be unable to tolerate the drought periods.

Poikilohydric and Homoiohydric Plants

Plants in general can be divided into two types depending on the water content of their cells. *Poikilohydric* plants have small cells with no central vacuoles.

Their water content depends on the humidity of the surroundings, and as they dry out their protoplasm shrinks in a regular manner, thus retaining its fine structure, although all life processes cease. As the relative humidity of the air increases the protoplasm reabsorbs water and life processes begin again. Poikilohydric plants include bacteria, blue–green algae, lichens, and fungi. The relative humidity required for active life processes varies with species, some soil bacteria requiring at least 95 per cent, while some fungi can grow when the relative humidity is only 60 per cent. The first plants to colonize the earth were poikilohydric and this characteristic has been retained in higher plants in pollen grains and seed embryos.

Poikilohydric plants eventually evolved into *homoiohydric* forms in which the cells have a large, central vacuole for the storage of water. The presence of such water helps to compensate for any reduction in humidity, but it is soon lost. On dehydration the protoplasm of homoiohydric plants loses its fine structure, as it is unable to shrink at a regular rate. Such protoplasm cannot be reactivated and thus the plant dies. The ancestors of the homoiohydric plants, the green algae, must therefore live in permanently moist habitats. However, more advanced plants have evolved a thick coating, called the *cuticle*, which is highly efficient in reducing the loss of water from their cells when the relative humidity is low. It was not until homoiohydric plants with a cuticle evolved that plants could move away from permanently wet situations and spread throughout the earth, even colonizing deserts.

Evaporative Power of the Air

The atmosphere is nearly always sufficiently dry to allow evaporation from the surfaces of plants and the soil. The *evaporative power* of the air is the amount of water which can be so absorbed and it is increased by high temperatures and wind, as well as by a low relative humidity. The evaporative power of the air is important to plants in two ways – it affects transpiration and influences soil moisture. When the sun shines on a wet soil the latter loses water by evaporation more rapidly than a water surface, such as a lake. This is because the heat input is concentrated at the surface and because minute irregularities of the soil increase the surface area exposed to the air. Surface evaporation can desiccate a normal soil to some depth. Thus water which cannot penetrate quickly is only available to plants for a short time.

Transpiration

Primitive plants originated in the seas where protoplasm evolved to a high state of complexity under water-saturated conditions. Thus, when plants began to colonize the land, where water was not so abundant, they had to devise means of maintaining the high minimum amount of water required by protoplasm for life processes to take place. In the terrestrial environment plants must endure an almost continuous loss of water to the atmosphere, therefore landward migration became possible only as efficient adaptations to the new conditions were developed.

We have already seen that land plants with two different types of cells occur. Poikilohydric plants, whose cells have no central vacuole, are able to survive dry periods by becoming completely inactive. Homoiohydric plants, however, are able to stay active when the humidity is low, as their cells have large vacuoles in which water can be stored. However, the homoiohydric group, to which most higher plants belong, must have a continuous supply of water to replace that which is lost to the air. Over 98 per cent of water absorbed by plants is lost to the atmosphere through the process known as *transpiration*.

The reasons why plants must transpire and thus lose water are twofold. Although the aerial surface of most plants is covered with a thick, almost waterproof coating, the cuticle, this coating must contain openings, known as the *stomata*, through which gases can pass. Oxygen needed for respiration and carbon dioxide for photosynthesis pass into the plant through the stomata, while the waste gases, carbon dioxide from respiration and oxygen from photosynthesis, pass out. While the stomata are open water is also drawn out of the plant. Transpiration takes place even in the humid tropics as the surface of the leaf is often several degrees hotter than the surrounding air, due to radiation from the sun. Thus a water vapour pressure gradient is built up which allows plants to transpire freely. In fact, the continual flow of water upwards through a plant, known as the *transpiration stream*, is the only means by which mineral nutrients can be transported from the roots to the growing shoots. Leaves exposed directly to the sun become overheated and die unless they are cooled. The evaporation of water from leaf surfaces requires heat, and thus transpiration is an important cooling mechanism.

The loss of water through the cuticle cannot be regulated by the plant, but as less than 5 per cent is transpired in this way it is of little importance compared to that lost through the stomata. However, the opening of the stomata, which is controlled by guard cells, depends on other factors besides the water needs of the plant and these are discussed in Chapter 5. The stomata of most plants close at night.

The rate of transpiration from a leaf is dependent on a number of factors, of which the evaporative power of the air is only one. These other factors include the temperature difference between the leaf surface and the air; the water content of the leaf tissues; the response of the guard cells to light and hence the opening of the stomata, and the action of light in increasing the permeability of protoplasm. It can be seen that all these factors are dependent on solar radiation and they are discussed in more detail in Chapter 4.

Evapotranspiration

Water enters the atmosphere by the processes of evaporation and transpiration. Evaporation takes place from water surfaces, soil surfaces, and plant surfaces, such as leaves, which intercept rain. The combined processes of evaporation and transpiration are known as *evapotranspiration* (fig. 3.4). The rate of evapotranspiration depends on temperature, relative humidity, wind speed, and longwave radiation.

The Water Balance of Plants

In order to transpire freely a plant must obtain continual supplies of water, and most plants colonizing the land have developed roots which can penetrate far down into the soil to search for, and absorb, water. Thus a plant can be pictured as a medium through which water passes from the soil to the atmosphere (fig. 3.5). This medium, however, is a living organism, and thus the amount of water retained within the plant is of vital importance. The actual state of hydration is known as the *water balance* of the plant, and can be defined as:

$$W_B = A - T$$

where W_B = water balance
A = water absorbed by the plant
T = water transpired by the plant

The water balance of a plant is determined by both internal and external factors. The external aspects consist of the amount of water available to the absorbing organs and the intensity of the transpiration promoting factors. The internal aspects depend largely on the structural and functional characteristics of the plant which tend to offset or aggravate the lack of water in the environment.

Many studies have shown that the absorbing and conducting systems of terrestrial plants are relatively inefficient in supplying water to meet the de-

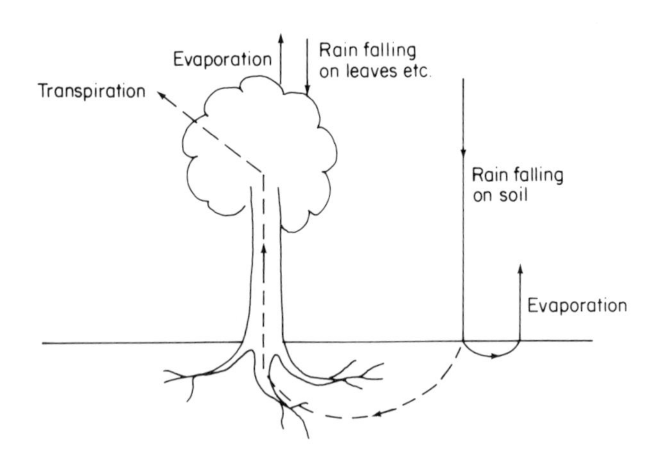

Fig. 3.4
Evapotranspiration

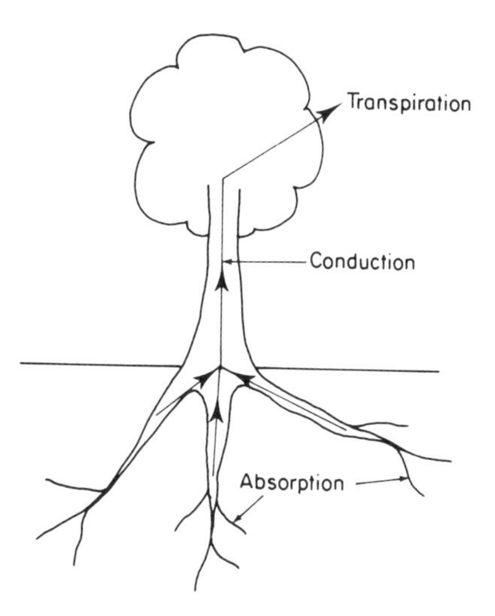

Fig. 3.5
Water movement in plants

mands of transpiration, even when growth water is plentiful. The tissues between the epidermis and xylem of the root can only transmit water at a slow rate, while the xylem in many plants is not very efficient in conducting water. Although many conifers grow in wet habitats they have developed thick cuticles, leaves with small surface areas and sunken stomata to overcome the very slow conduction of water by the xylem of the roots of these trees. Even under conditions of high temperature conifers show a much lower rate of transpiration than most other trees.

Even when the soil contains ample water, most plants lose more during the hours of daylight than they are able to absorb and conduct. Thus their water balance is negative. This can result in a loss of as much as 40 per cent of their normal water content, causing cells, especially those of the leaves, to shrink and thus decrease the area and thickness of the leaf blade. Further loss of water leads to such a reduction in turgor that stems and leaves wilt.

During the night, however, the stomata in most plants close so that transpiration is reduced to 3–5 per cent of that during the daytime. Water absorption remains constant and thus the water balance becomes positive. In fact so much water may be absorbed that it is deposited on the surface of the leaves as water droplets, a process known as guttation.

The structure of plants appears to be more strongly influenced by water balance conditions under which they are grown than by any other single factor of the environment. In comparison with plants grown under conditions of ample water, those grown under conditions of drought have the following characteristics:

1. Reduced shoot size (nanism).

2. Increased size of root system.
3. Smaller cells in leaves, which result in smaller and thicker leaf blades and stomata closer together.
4. Thicker cuticle and cell walls.
5. Smaller intercellular spaces.
6. Smaller xylem cells.

Plants with the above structural characteristics are termed *xeromorphic*.

Susceptibility to drought varies throughout the life cycle of the average plant. Seeds are often capable of enduring extreme drought, as in deserts, and, indeed, may require such a condition to maintain their viability. However, after the first leaves unfold, seedlings usually become very susceptible to drought and are easily desiccated.

As plants approach maturity they become less sensitive to lack of water, although reactions to drought vary with species. For example, sorghum and maize, which are similar morphologically, react quite differently to lack of water during the growing season. Sorghum suspends development with no serious subsequent ill effects, until water is again available. Maize, however, continues to develop so that pollination and other functions are seriously affected.

Temporary and Permanent Wilting

Temporary wilting is a common phenomenon in the tropics when the water balance of plants becomes negative during the day, due to the power of the sun. However, as long as the soil contains water available to the roots, the water balance is restored during the night and the plants do not die. In an attempt to correct the water balance the stomata of a wilted leaf close. This prevents water loss but also impedes the absorption of carbon dioxide. Thus photosynthesis ceases and growth is retarded.

In order to prevent or reduce temporary wilting of crop plants such as coffee or cocoa, the planting of shade trees is common in the tropics. Such trees intercept some of the radiation from the sun, so that transpiration of the crop plants is reduced.

If insufficient water is available in the soil to restore the water balance during the night and a plant is continually exposed to strong sunlight during the day, *permanent wilting* eventually occurs. Under such conditions the tissues become so damaged that they are unable to recover even if water does become available. Woody plants and annuals usually die within two weeks, but perennial herbs become dormant, their aerial parts dying but the root systems remaining alive for a considerable time. At the onset of rain the aerial shoot system begins to grow again. Grasses are particularly good at surviving permanently wilting conditions, but in those regions of the tropics with a long dry season many other perennials are able to survive from one rainy season to the next in this manner.

Soil Moisture

The state of water within the soil is very complex, varying from free flow to firm adsorption on the surface of soil particles. Three main types of soil water occur, although there is no clear distinction between the types, as each merges into the other. These states are known as gravitational water, capillary water, and hygroscopic water (fig. 3.6). Water is also held as water of hydration when it is chemically combined in the hydrated oxides of iron, aluminium, silicon, etc. This water is very difficult to remove and is thus completely unavailable to plants. Soil water is discussed in more detail in Chapter 2.

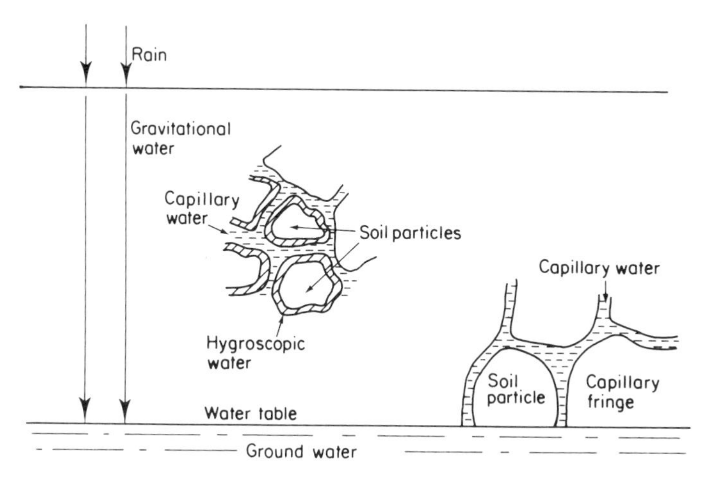

Fig. 3.6
Soil water types

Soil Moisture Constants

The actual water content of a soil is variable, but such characteristics as the maximum amount of capillary or hygroscopic water a particular soil can hold are determined mainly by the nature of the soil particles and are therefore constants. Important soil constants include the field capacity and the permanent wilting percentage.

Field capacity

The *field capacity* is the amount of water remaining in the soil after the gravitational water has drained away but before any other losses occur. It is therefore the maximum amount of water that can be held as films and in the pores of soil which is out of contact with the water table.

As long as there is water in excess of field capacity the surface of the soil remains moist, because capillary water replaces water lost by evaporation.

When the moisture in the soil drops to, or below, field capacity it can no longer move upwards and roots must continually branch out to search for the water needed by the plant. For this reason the roots of plants grown in soils in which the moisture content frequently drops below field capacity tend to be more branched than those of plants in wetter soils.

Field capacity is also of great ecological importance as a measure of the maximum storage capacity of the soil, and its determination is required in order to calculate the amount of irrigation water which will wet a soil to the depth of root penetration.

Permanent wilting percentage

The roots of a plant growing in moist soil absorb water which is then transpired through the leaves to the atmosphere. Thus there is a steady reduction in soil moisture and, if no further rain falls, a point is reached where the rate of absorption is less than that of transpiration. As water is drawn out of the plant but not replaced the turgidity of the cells decreases and wilting occurs.

The earliest, or temporary, wilting may occur only during the hottest part of the day, the plants regaining their turgidity during the night. However, a point is eventually reached when a wilted plant will not revive even though the air around its leaves is at 100 per cent relative humidity and transpiration ceases. This is the *permanent wilting point* and the amount of water remaining in the soil at this point is the *permanent wilting percentage* or *wilting coefficient*. This coefficient varies with the nature of the soil and includes all the hygroscopic water and the capillary water which is held too strongly for easy absorption by the roots. Clay soils with their high surface areas are able to hold much unavailable water and thus have higher wilting coefficients than sandy soils.

Since growth does not usually cease until the moisture content of a soil has been reduced to the permanent wilting percentage, water in excess of this value is known as the *growth water* or *available water*. The latter term is not strictly true, however, as dormant plants can extract small amounts of water below the permanent wilting percentage, but only at a very slow rate. From an ecological point of view a soil with a moisture content below the permanent wilting percentage is a dry soil, regardless of the actual water content.

By subtracting the permanent wilting percentage from the field capacity the amount of water available to all plants growing in a particular soil can be determined:

$$W_g = F - P$$

where W_g = growth water
F = field capacity
P = permanent wilting percentage

The drying out or desiccation of a soil begins at the surface and gradually extends downwards. Thus surface rooted plants, such as mosses, are the first to cease growth, followed by shallow-rooted mature plants and seedlings of deeper-rooted plants. The last to cease growth or die are those most deeply rooted, and even in arid regions there are plants with roots in permanent contact with the capillary fringe (fig. 3.7).

Plants, such as annuals, with no water storage capacity die within a few days, or even hours, after the moisture content of the soil reaches the permanent wilting percentage. Succulents, however, are able to store water which is used to replace transpiration losses for some time after soil moisture has fallen below the permanent wilting point. Some perennials, especially certain grasses, are so able to reduce their transpiration rates at the permanent wilting point that loss of water is almost equal to the very slow rate of absorption.

Investigation of many species of plants has shown that the permanent wilting percentage is a characteristic of the soil, rather than of the plants growing in the soil. Thus desert plants experience similar permanent wilting percentages to those of wetter regions. The amount of residual moisture remaining in the soil at the permanent wilting point is determined by an abrupt change in the physical forces controlling the availability of water in thin films, which most plants are unable to overcome. However, some xerophytes are able to extract water from a soil which is below the permanent wilting point, due to the high concentration of solutes in their cell solutions.

Many plants have evolved methods of overcoming the effects of a temporary drought when the soil moisture falls below the permanent wilting percentage, but few can survive prolonged drought. Some species of algae have been found living after 50 years of drought, while the seeds of desert plants can usually remain viable until rain falls, even though this may not be for several years.

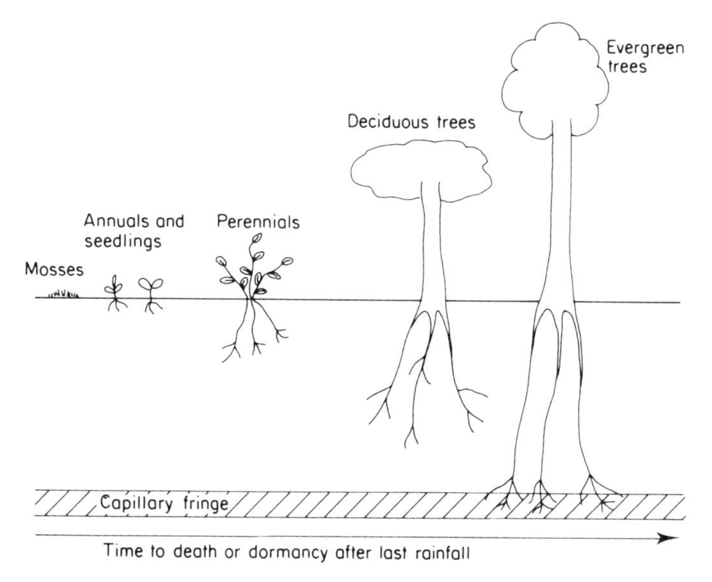

Fig. 3.7
The effect of root length on the ability of plants to overcome drought

Gravitational Water

After a heavy shower of rain or artificial watering the surface layer of the soil is temporarily saturated. Due to the force of gravity this water rapidly descends through the dry upper layers leaving a zone of moist soil as it passes (fig. 3.6). If sufficient water has fallen on the surface this *gravitational water* eventually reaches the permanently water saturated soil known as the *water table*. All water below the water table is termed the *ground water*. The water table usually lies just above the bedrock, but perched or hanging water tables also occur above impervious layers such as the iron pans discussed in Chapter 2.

When the gravitational water reaches the water table it ceases to move vertically and joins the ground water in moving horizontally until it emerges at a lower surface to collect in ponds, lakes, etc. or to join streams and rivers.

The water table is seldom level for any distance, one factor influencing its height being the vegetation cover on the surface. Dense vegetation with high transpiration rates has the effect of lowering the water table. Thus trees are often planted to decrease the swampiness of a region. Conversely the presence of trees in an arid region may result in a water table so deep that shallower rooted vegetation is unable to survive.

Capillary Water

Capillary water is very important to plants, as it is the source of almost all the water extracted by the roots from the soil. As the gravitational water drains downwards it leaves behind much moisture in the pores between the particles of soil. This water does not move downwards, but due to the forces of capillarity it can travel upwards against the pull of gravity.

The rate of movement of capillary water through the soil depends on the structure, texture and temperature of the soil, and is particularly influenced by the thickness of water films within the pores. Thick films can move in any direction but thin films are held so strongly by the soil particles that they are immobile. Thus the amount of capillary water available to plants depends on the thickness of the films. The texture of the soil is also important, as fine textured soils have more pore spaces than coarse soils, while the smaller pore size of the former restricts the movement of water.

The amount of lateral movement of capillary water is not great so that irrigation water must be applied close to the base of shoots.

There is a permanently moist region of soil above the water table, known as the *capillary fringe*, in which the pores are filled with water drawn upwards from the ground water. However, as the distance from the water table increases the forces of capillarity become weaker so that eventually the capillary fringe ends. As the capillary fringe is usually well below the surface of the soil, evaporation has no effect on its moisture content. However, if the water table is near the surface much moisture can be lost from the capillary fringe by direct evaporation.

In arid climates many plants are able to utilize the water in the capillary fringe. These plants usually have tap roots which can penetrate the soil as far as this permanent water, provided the water table is not too deep. However, few plants can grow with their roots below the water table, due to lack of air within the soil pores, which are completely filled with water. The depth to which the roots can penetrate the capillary fringe depends on the aeration requirements of a particular plant.

Plants depending on the capillary fringe for their water supply must be long lived and produce seeds every year, as there will be few years in which sufficient rain falls to maintain the seedlings until their roots are long enough to reach the capillary fringe. Most seedlings germinate but later die through lack of water. Alfalfa, for example, can only be grown in arid regions if the seedlings are irrigated until their roots have penetrated the capillary fringe, after which the field can be cropped repeatedly without further irrigation.

Many plants growing along stream banks in otherwise arid regions depend on capillary fringe water. Such plants usually have high transpiration rates and can thus be responsible for a stream drying out more rapidly than it otherwise would during a drought. However, the usefulness of such plants in preventing erosion and floods probably outweighs their drying effects.

Hygroscopic Water

Evaporation from the soil surface and absorption by the roots of plants gradually reduces capillary water. Eventually a state is reached where the water molecules are held so tightly by the soil particles that they can only be removed under conditions of severe drought. This strongly held water is known as *hygroscopic water*, and in general little is available to plants. Hygroscopic water exists in equilibrium with the vapour pressure of water in the soil atmosphere and therefore forms its thickest coat around soil particles when the relative humidity is 100 per cent.

Water Vapour

The portion of pore space not filled with water contains a soil atmosphere, and as long as capillary water is present in the soil this atmosphere will be almost saturated with water vapour. When capillary movement of water ceases, small amounts of moisture are transferred through the soil by vapourization from thick films and condensation where the films are thinner. The vapour flow depends on the temperature and humidity and much water vapour is lost from the surfaces of soils exposed to the sun.

Waterlogged Soils

Plants can suffer from too much water in the soil, as well as too little, although it is not the overabundance of water that is the problem but lack of air in the soil pores. Because the absorption of water by the roots is closely linked to the transpiration of water through the leaves, plants seldom absorb too much water. Should this happen the excess is deposited on the leaves as small drops of liquid, by the process of guttation.

Marshes and swamps in which the soil is always saturated with water are world-wide. They result from permanent or seasonal flooding, the soil never drying out to field capacity. In a normal soil many of the pores are filled with air, so that diffusion of the gases oxygen and carbon dioxide is rapid. In waterlogged soils, however, the pores are filled with water so that gases have to diffuse through the soil solution, which is a very slow process. The diffusion of oxygen from the outside atmosphere is too slow to maintain aerobic organisms, so that waterlogged soils contain an abundance of anaerobic organisms which use nitrates or sulphates as their source of oxygen for respiration. Anaerobic conditions also prevail below and just above the water tables of normal soils.

The reduction of nitrates, a process known as *denitrification*, may produce nitrogen gas, a compound which has no nutritional value for higher plants. Sulphate reduction results in the formation of hydrogen sulphide, which is a plant toxin. Ferric compounds are reduced to ferrous and thus in a soil high in iron compounds hydrogen sulphide is precipitated as ferrous sulphide.

Despite the lack of oxygen and the often low nutritional status of waterlogged soils, many plants are adapted to such an environment, the most important to man being rice (Plate 3.3). Rice and other wetland plants have large intercellular air spaces which enable them to transport oxygen from the shoots to the roots. Some of this oxygen diffuses into the soil, so that an aerobic layer is formed around the roots. Aerobic organisms living in this rhizosphere oxidize ferrous ions to ferric and manganous ions to manganic, thus removing potential toxins from the vicinity of the root. The *Beggiatoa* bacterium protects rice against the toxicity of hydrogen sulphide by oxidizing the sulphide ion to sulphur. In order to do this the bacterium needs an external supply of the enzyme catalase and this is provided by the rice roots. Thus the association of rice and *Beggiatoa* is an example of the symbiosis between plants described in Chapter 7.

The Classification of Plants According to their Water Requirements

Plants can be classified according to their water requirements, as follows:

Hydrophytes – plants that live in water
Helophytes – plants that grow in waterlogged soils
Mesophytes – plants that will neither tolerate too little nor too much water
Xerophytes – plants adapted to dry habitats

The hydrophytes can be divided into four types: floating hydrophytes, suspended hydrophytes, submerged, anchored hydrophytes, and floating-leaved, anchored hydrophytes. The floating hydrophytes are in contact with both air and water but not with soil. They include the water hyacinth (*Eichhornia*), a very troublesome water weed of the tropics. The suspended

Plate 3.3
Rice field. (Reproduced by permission of A.A.A. photo, Paris.)

hydrophytes, such as the duckweeds (*Lemna*), are only in contact with water and can travel large distances when caught in river currents. Submerged, anchored hydrophytes have their roots in soil but are out of contact with the air. The pondweeds (*Potamogeton*) belong to this group. Totally submerged plants contain large intercellular spaces (lacunae), which act as gas storage chambers. These spaces are alternately filled with carbon dioxide from respiration and oxygen from photosynthesis.

The floating-leaved, anchored hydrophytes, such as the water lilies (*Nymphaea*), are in contact with air and soil as well as water. The leaves of such plants are usually buoyant and float above the water, so that absorption of oxygen and carbon dioxide can easily take place. The buoyancy results from large air spaces or lacunae between the cells. The lack of oxygen in the vicinity of the roots of these plants is not detrimental as the gas is swiftly and efficiently transported to all parts of the plant from the leaves. In order that gas absorption should be unimpeded the leaves of floating aquatics are either waxy, so that water droplets are not held, or hairy, so that droplets are held above the surface leaving an air space. The famous water lily *Victoria amazonica* has leaves with many small perforations through which water droplets can drain.

Floating islands, consisting of mats of vegetation, are characteristic of tropical waters. Species commonly found in such islands include many reeds (Cyperaceae), *Ipomoea reptans* and *Polygonum barbatum*, as well as some grasses.

The helophytes grow in shallow water and waterlogged soils and include such diverse species as rice (*Oriza sativa*) and mangroves (*Rhizophora* and *Avicennia*).

Neither hydrophytes nor helophytes suffer from lack of water, but many mesophytes must be able to withstand short periods of drought. Annuals overcome the problem of water deficit by completing their life cycle before the soil dries out to permanent wilting point. They then pass the period of drought as seeds, whose water requirements are very low. Many perennials reduce their rate of transpiration considerably by losing their leaves during periods of drought and becoming dormant. During dormancy, water, nutrient, and gas requirements are very low, being needed only in sufficient amounts to keep the plant alive. The geophytes, many of which are members of the Araceae and Liliaceae families, have underground swollen parts known variously as tubers, bulbs, corms, rhizomes, etc., in which water absorbed during the wet season can be stored for use during drought. Tropical evergreen rain forest trees are mesophytes which seldom suffer from lack of moisture. Thus they have no need for special adaptations to overcome the problems of drought.

There is no sharp dividing line between mesophytes which can tolerate a few months of drought and the xerophytes, the most highly adapted of which can survive several years of drought. In general, there are three main ways in which xerophytes maintain their water balance despite the aridity of their surroundings – reduction of water loss, increase in water absorption, or conservation of water within the tissues of the plant. Water loss can be minimized in a number of ways, many desert plants having very small leaves (microphylls), or leaves reduced to spines or scales, as in members of the Cactaceae family. The leaves of many *Acacia* species of savanna regions are reduced to very small leaflets or are absent altogether, photosynthesis being carried out by the green petioles. As xerophytes usually grow in areas of high light intensity photosynthesis is not affected by the smallness of the leaves. Plants growing in more shaded habitats must have larger leaves, but these are covered by a thick, waxy coating (sclerophylls) which prevents cuticular loss of water. The stomata are often sunk in pits below the surface of the leaf, so that despite the low humidity of the atmosphere a layer of humid air builds up above the opening, reducing the rate of transpiration. The leaves of most evergreen savanna trees are similarly constructed. Hairy leaves achieve the same effect, while the rolling or folding of leaves, characteristic of many grasses, reduces the surface area from which water is transpired. Plants growing in arid regions usually close their stomata during the hottest part of the day, while many succulents, particularly those belonging to the Crassulaceae, only open their stomata at night. When water is available, however, xerophytes can transpire at rates equalling those of mesophytes.

A high absorption of water by the roots of xerophytes is achieved by an increase in the volume and rate of root growth compared to that of most mesophytes. The extensiveness of the root systems in proportion to the shoot systems in some xerophytes is an important adaptation, as not only does it increase water absorption but also ensures that only a relatively small proportion of the plant is exposed to the atmosphere.

Some tap roots reach the capillary fringe and the plant can thereafter transpire freely. Such plants are not true xerophytes as they have avoided the dryness of their habitat, rather than becoming adapted to it.

A large number of xerophytes, known as *succulents*, are able to store water within their tissues. The succulent parts of such plants contain a large number of cells whose vacuoles have expanded at the expense of the intercellular spaces. After a fall of rain these vacuoles become full of water and the succulent organs swell considerably. As the water is depleted during a drought the organs gradually shrink. In this way the plant is able to exist from one wet period to another, even though these may be separated by several months.

Succulence may occur in roots (*Ceiba parvifolia*), stems (Cactaceae, Euphorbiaceae) or leaves (Agavaceae). Succulent plants are probably those most highly adapted to survive a prolonged period of drought. They have low transpiration rates, and further conserve water by closing their stomata during daylight. Such plants have a special method of fixing carbon dioxide for photosynthesis, described in Chapter 4. The cacti are unique amongst desert plants for their shallow root systems. Rain which wets only the surface layer of the soil is of direct benefit to such plants, deeper rooted xerophytes benefiting only from the increased humidity. In some cacti fine rootlets are produced soon after a shower of rain, which die as soon as the soil dries out again.

Halophytes

Plants growing in saline soils are well adapted to coping with the stresses caused by high concentrations of ions. Such plants are termed *halophytes* and many are members of the Chenopodiaceae family. The effects of high concentrations of salts on plants are partly physical, due to osmotic pressure, and partly chemical, due to the nature of the ions. In order to maintain the water balance of normal plants (glycophytes) the concentration of dissolved solids in the soil solution must be less than that within the root cells. Salt tolerant plants maintain this balance by decreasing their absorption of water. Halophytes, however, have a differnt mechanism and are able to absorb water freely. The cells of halophytes contain a much higher concentration of salts than those of glycophytes, so much so that they may taste salty. Halophytes also excrete salts onto the surface of their leaves, forming an incrustation. Many are succulent, producing tissue which can absorb and store much water. It has been found that succulence can be stimulated in non-halophytes by the presence of chloride ions.

In both glycophytes and halophytes germination and seedling survival is difficult under saline conditions. Thus under such conditions seeds of halophytes germinate in the wet season after the salts have been washed out of the surface layers of the soil. In some plants, such as the mangrove, the seed germinates before it leaves the parent tree, a condition known as *vivipary*.

Water and Plant Growth

It is well known that lack of water causes a cessation of growth in plants, coupled with reduced photosynthetic activity. However, the role of water in promoting growth is far from clear. Many tropical trees periodically produce a rapid expansion of new leaves, a process known as *flushing*. These leaves are often red or pale coloured and are thus very conspicuous against the dark green of the older leaves. In some trees flushing occurs during the latter part of the dry season, while in others, such as teak (*Tectona grandis*) and *Terminalia superba* flushing seems to be the result of increased water content in the soil. Day length and temperature are also important aspects in the control of flushing (see Chapter 4).

Leaf-fall in deciduous trees is a direct consequence of water stress due to drought, the trees passing most of the dry season in a leafless state which conserves moisture. The production of new leaves at the end of the dry season enables the tree to take full advantage of the increased water content of the soil when the rains begin, photosynthesis then taking place of maximum rate.

Most tropical plants have a definite flowering time, but some evergreen trees flower almost continuously. On others, some branches may flower while other branches are fruiting. Many deciduous trees flower during the dry season. Flowers do not have stomata and so little water is lost through their surfaces.

The flowering of geophytes (plants with bulbs, corms, or tubers under ground), especially those belonging to the Amaryllidaceae family, is due to a sudden increase in soil moisture. Such plants usually all flower together over a large area, about three days after a heavy shower, a process known as *gregarious flowering*. The immature flower buds are held within the bulb until conditions are right for their emergence. Gregarious flowering gives a species a definite advantage as it ensures cross pollination can easily take place.

A coffee plantation will also suddenly burst into bloom about ten days after heavy rain, but here the factors inducing flowering are more complicated. A plantation which is continuously irrigated will not produce flowers, the bushes needing at least three weeks of drought before rain falls. However, it has been noted that a sudden drop in temperature, such as usually occurs in the tropics when rain falls after a drought, will initiate blooming, even in irrigated crops.

The germination of seeds is a complicated process depending on several factors, of which water is only one. Many cacti seeds need to be soaked with water to remove growth inhibitors before they will germinate, a process which is probably necessary for the seeds of many other species.

Water is the medium of dissemination of some tropical fruits and also of the pollen of aquatic plants, while the sperms of lower plants must, in general, swim to reach the egg. The dissemination or transportation of fruits by water is only common amongst hydrophytes as the seeds of mesophytes and xerophytes are easily killed by lack of oxygen when soaked. The fruits of the coconut (*Cocos nucifera*) can endure long periods of submersion, however.

Suggestions for Further Reading

Etherington, J. R. (1975). *Environment and Plant Ecology*. Wiley.
Larcher, W. (1980). *Physiological Plant Ecology*. Springer-Verlag.
Levitt, J. (1980). *Responses of Plants to Environmental Stresses* (second edition). Academic Press.
Longman, K. A. and Jenik, J. (1974). *Tropical Forest and its Environment*. Longman.
Richards, P.W. (1952). *Tropical Rain Forest*. Cambridge University Press.
Sutcliffe, J. F. (1979). *Plants and Water* (second edition). Arnold.
Tivy, J. (1971). *Biogeography*. Oliver and Boyd.
Treshow, M. (1970). *Environment and Plant Response*. McGraw-Hill.
Turner, N. C. and Kramer, P.J. (1980). *Adaptation of Plants to Water and High Temperature Stress*. Wiley.
Villiers, T. A. (1975). *Dormancy and the Survival of Plants*. Arnold.
Walter, H. (1971). *Ecology of Tropical and Sub-tropical Vegetation*. Oliver and Boyd.

Chapter 4

Plants and Radiation

Introduction

All life on earth ultimately depends on the energy contained in radiation from the sun for its continued existence. Plants obtain this energy directly, but animals have to depend on chemical energy locked up in carbohydrates synthesized by plants for their requirements.

Radiant energy from the sun is received by the earth in the form of electromagnetic waves, which vary in length from about 5000 to 290 nm. Waves with lengths shorter than 290 nm are completely absorbed before they reach the earth's surface. The solar spectrum can be divided into three main regions (fig. 4.1), of which the band containing waves of lengths 750–400 nm is the most important to plants and animals. This band of waves, known as *visible light*, is able to pass through the earth's atmosphere with little loss of energy, although much reflection due to water vapour and particles in the air may occur. Light scattered by water droplets and gas molecules becomes diffuse light or sky light, as contrasted with direct light from the sun. On a clear day diffuse light may only comprise 10–15 per cent of total light, but on an overcast day it can reach 100 per cent.

The angle of the sun's rays with respect to the earth's surface at any given point determines the distance these rays must travel through the atmosphere and hence the percentage of absorption of energy that takes place (fig. 4.2). Latitudinal variations in light intensity due to the height of the sun above the horizon are very important. In equatorial regions the light is most intense and as much as 70 per cent of direct sunlight may reach the earth's surface. In

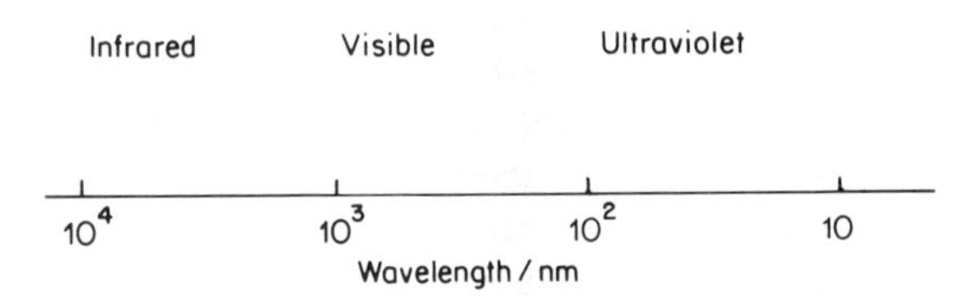

Fig. 4.1
Electromagnetic waves received by the earth's surface

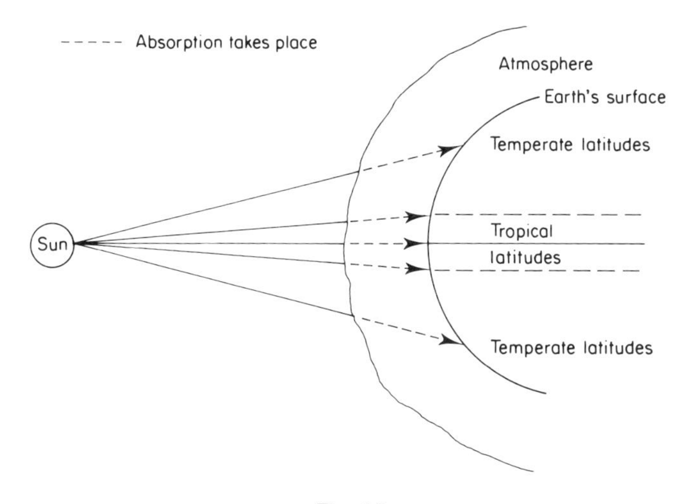

Fig. 4.2
Absorption of energy by the earth's atmosphere

temperate regions, however, only around 50 per cent direct sunlight reaches the surface, even on a clear day, while the proportion of diffuse light is higher than that at the equator.

Electromagnetic waves with lengths shorter than 390 nm are known as *ultra-violet light* (meaning 'above violet'). These waves have only minor effects on plants, as under natural conditions they are mostly absorbed by ozone and oxygen molecules in the atmosphere. These gases screen the earth's surface from the harmful effects of ultraviolet radiation, and any rays which do reach the surfaces of plants are absorbed by the epidermis, so they have little effect on metabolic processes. Some flowers reflect ultraviolet light as many insects 'see' in this range, although the human eye is not so adapted.

Radiation with wavelengths longer than 750 nm is known as *infrared light* ('below red'). That received from the sun has a maximum wavelength of 3000 nm and is known as near infrared radiation. Much of this radiation is absorbed by carbon dioxide and water vapour molecules in the atmosphere. This screening effect is essential to plants, for if they were exposed to all incoming infrared radiation they would become too hot for life processes to continue. The surfaces of leaves also help to reduce the absorption of this radiation, as they can reflect as much as 70 per cent.

Eventually all solar radiation is converted to waves with lengths exceeding 3000 nm, known as the far infrared, and reflected back through the atmosphere into space. The heating effects of the far infrared range are very important to plants, while the near infrared exerts control over some physiological processes.

The earth's surface acts as a buffer between the radiation extremes of day and night. During the day it receives energy from the sun (*insolation*) and warms up, while at night it radiates heat and slowly cools down. Without this buffering effect most plants would find the temperature extremes of day and night intolerable.

During their evolution, plants have developed systems, collectively known as *photosynthesis*, by which they are able to convert the energy contained in some visible light waves to chemical energy. However, although this chemical energy must maintain the whole biomass, it is derived from only a small proportion of the radiant energy absorbed by plants. Most of the absorbed energy is converted to heat and used to evaporate water in the transpiration process described in Chapter 3. Some heat is used to warm the plant, while the remainder is passed to the surroundings, mainly by radiation, but also by convection and conduction if the air is at a lower temperature than the plant.

Soil Temperature

As the sun rises the earth's surface begins to gain more heat than it loses by conduction and reradiation, so that its temperature rises rapidly. After several hours a high surface temperature is reached and this is maintained during the greater part of the day, radiation gains being approximately equalled by losses. After sunset the earth's temperature declines slowly, loss of heat being accelerated by the cooling effect of evaporation from the soil. Soil temperatures characteristically drop below air temperatures with the minimum occurring just before sunrise. Because the daily maxima are higher and the nightly minima lower, the surface temperature of exposed soil fluctuates more widely each 24 hours than does the air temperature.

Although the temperature of the surface layers of soil fluctuates, in the tropics the temperature of the soil about 1 m below the surface approximates to the average annual air temperature and remains almost constant throughout the year.

The colour of a soil surface affects the amount of radiation it can absorb. White reflects all radiation, while black absorbs it completely. When a bare, light-coloured soil receives radiation from the sun, reflection is so strong that the air just above the soil surface becomes very hot, but the soil itself remains fairly cool. A dark surface, however, such as a burned area, absorbs radiation and can become relatively hot. It is not uncommon for adjacent light- and dark-coloured soil surfaces to differ in temperature by as much as 20°C.

Energy Absorption by Plants

The amount of the sun's energy absorbed by plants depends on many factors. Plants growing in shaded areas are exposed to less radiation than those in direct light. The orientation of the leaves is also important to energy absorption, as those at right angles to the sun's rays absorb much more than those in a vertical position (fig. 4.3). Some plants are able to change the orientation of their leaves to obtain maximum benefit from the available radiation.

Pigmentation also affects the amount of energy absorbed. Dark green leaves

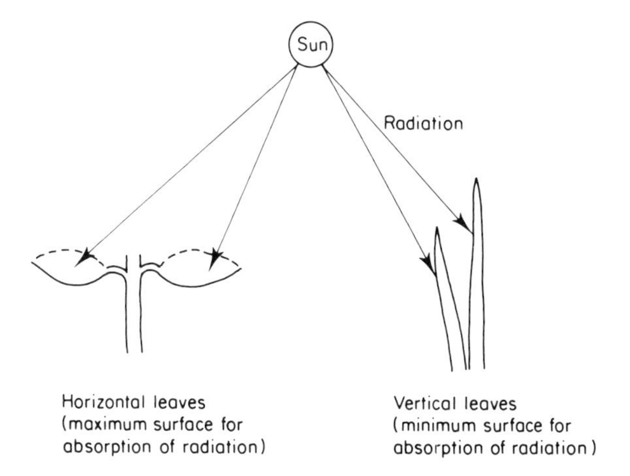

Fig. 4.3
The effect of leaf orientation on energy absorption

with abundant chlorophyll are able to absorb much more of the light required for photosynthesis than yellow leaves with little chlorophyll. Other pigments such as phytochrome, carotenoids, and flavonoids, especially anthocyanins, are also responsible for energy absorption of various wavelengths. The anthocyanin pigmentation of some tropical forest herbs enables these shaded plants to absorb more of the available light than leaves without such pigmentation.

Plants normally growing in places exposed to much direct sunlight usually have leaves with highly reflective surfaces, so that a high proportion of the energy received is reflected back into the surroundings. Such leaves can be shiny or have a dense covering of white hairs.

Plant Temperatures

Although plants are *poikilothermic*, meaning their temperature approaches that of their surroundings, there can be considerable differences between the temperature of a leaf surface and that of the surrounding air. Usually plant temperatures are higher than those of the air during the day and cooler at night. During the day, when still air is at a temperature of 35°C leaves can be at a temperature of 40–50°C. However, a passing cloud or breeze will exert a considerable cooling effect. Transpiration also cools and can reduce the temperature of leaves by 5–10°C.

In a stand of plants large differences between the temperature of leaf surfaces and that of the surrounding air only occurs in a narrow zone near the upper surface exposed to direct light. This zone is known as the *active layer* or *effective surface*. In a tropical forest the active layer would contain the tops of trees and their associated epiphytes, while in grassland savanna it would contain the tops of the grasses (fig. 4.4). Within the active layer there can be considerable differences between the leaf temperature and that of the air,

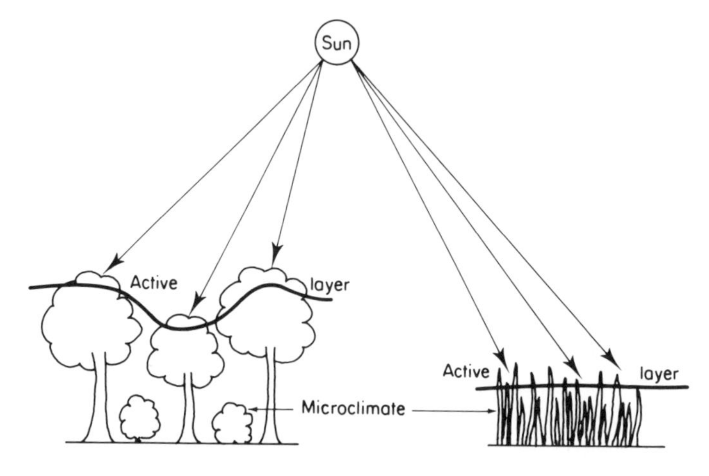

Fig. 4.4
The active layers and microclimates of forests and savannas

leaves becoming hotter by day and cooler by night. Below the active layer within stands of plants, however, there are much smaller differences in temperature. Differences in temperature and exposure to light have a considerable effect on the morphology of a plant, as will be discussed below.

Cardinal Temperatures

Most plants can only survive a narrow range of the known temperature scale, the term temperature denoting a particular level of molecular activity. The higher the temperature of a body the more its constituent molecules vibrate. There is little biological activity below 0°C or above 50°C because of two factors common to all living organisms – their high proportion of water, which freezes at 0°C, and the destruction of vital proteins above 50°C. Although there are some lower plants which flourish at temperatures below 0°C or higher than 50°C, most higher plants need an even narrower range if they are to thrive. For plants of temperate and cold regions this range lies at the lower end of the scale, such plants suffering from heat injury if subjected to temperatures above about 30°C. Tropical plants, however, seldom suffer from heat injury, but are unable to survive the lower extremes and some are fatally injured at temperatures which, although low, are well above freezing point.

The rate at which an enzyme catalyses a biochemical reaction usually depends on temperature and the temperature for maximum rate varies for different processes or for the same process in different plants. For each process there are three important temperatures, known as the *cardinal temperatures*. These are the minimum below which the process is not detectable, the maximum above which it is not detectable and the optimum at which the reaction progresses with maximum velocity. The minimum temperature for growth in such tropical plants as sorghum and the date palm lies between 15°C and 18°C, but for many plants of higher latitudes it lies between −2°C and 8°C. The maximum temperature for the growth of species of *Opuntia* was found to be near 56°C, but for many arctic plants this cardinal temperature is only a few degrees above freezing point. The cardinal temperatures of lower plants cover a much greater range than those of higher plants. There are algae whose minimum lies below 0°C and others, which live in hot springs, whose maximum is above 90°C.

Different functions of the same plant may have different cardinal temperatures. In most plants the optimum temperature for photosynthesis is distinctly lower than that for respiration. Because growth and reproduction depend on a more rapid rate of accumulation than of oxidation of organic compounds, plants are at a distinct disadvantage whenever the temperature rises above the optimum for photosynthesis. However, as we shall see, tropical plants have evolved ways of overcoming this problem. The photosynthesis–respiration relationship is important in setting the lower latitudinal and altitudinal limits for many plants, including apples, peaches, and potatoes.

Various organs of the same plant may have different cardinal temperatures for the same function. In general, roots have a lower minimum temperature than shoots. Cardinal temperatures also vary with the age of the plant, its physiological condition and with various environmental factors. Thus the term *cardinal temperature* covers a range of values, rather than referring to a fixed point.

Energy Flow

Energy flow is a one-way process from the sun to the earth. Radiation from the sun is continually received by the earth during the hours of daylight and is eventually lost to space. This does not mean that the energy of the sun's radiation is destroyed, only that it is spread out so thinly as to be of no use to life on earth. Thus for life on earth to continue, daily insolation is an absolute necessity.

The intensity of radiation received by the outer atmosphere is 1.39 kW per square metre of surface, a value known as the *solar constant*. On average, only about half of this radiation reaches the earth's surface, due to reflection and absorption by the atmosphere, especially of the long and short wavelengths. The intensity of direct visible light is little altered, however, unless clouds are present, a factor on which the whole of life depends.

There are large differences in the amount of radiation received by different places on earth, even within the tropics, as this depends on latitude, altitude, and the weather. The intensity of radiation decreases with increased latitude, increases with increasing altitude and is decreased by clouds and turbulence. During calm, cloudless days tropical regions receive the greatest amount of radiation – about 70 per cent of the solar constant at sea level.

Radiation falling on bare ground is partly used to warm the soil and partly reflected, light-coloured soils reflecting more and thus staying cooler than black soils, a factor which can influence the germination of some seeds.

However, much of the radiation reaching the earth's surface encounters plants of one kind or another. As it passes through a stand of plants it is absorbed and reflected so that only a few per cent reaches ground level. In a dense tropical forest, for instance, the amount reaching the floor can be less than 1 per cent.

Radiation falling on a plant is partly absorbed, partly reflected and partly transmitted, the relative proportions depending on the structure of the leaf (fig. 4.5). Shiny or hairy leaves reflect more radiation than smooth, dull leaves; thin leaves with little cuticle transmit more than thick leaves with a dense cuticle. Of the absorbed energy most is converted to heat, only a small proportion being stored as chemical energy for use by the whole biomass.

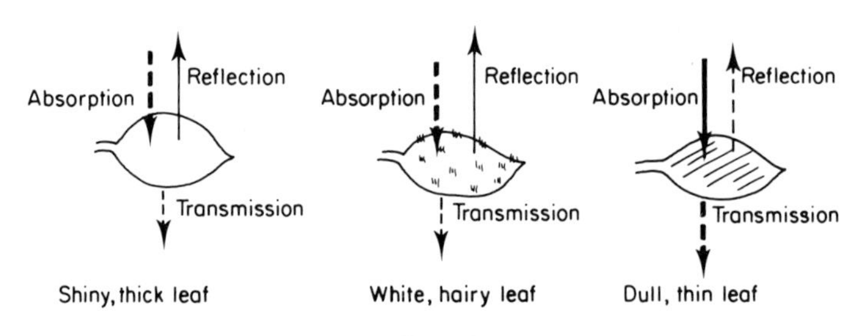

Fig. 4.5
The absorption, reflection, and transmission of light by various types of leaf

During the photosynthetic process, which converts solar energy to chemical energy, much of the absorbed radiation is lost as heat. Further losses of energy as heat also occur at every stage of the food chain. The passage of energy from the sun through plants, primary consumers, secondary consumers, and decomposers, until all is finally dispersed as heat can be summed up in the energy flow diagram shown in fig. 4.6.

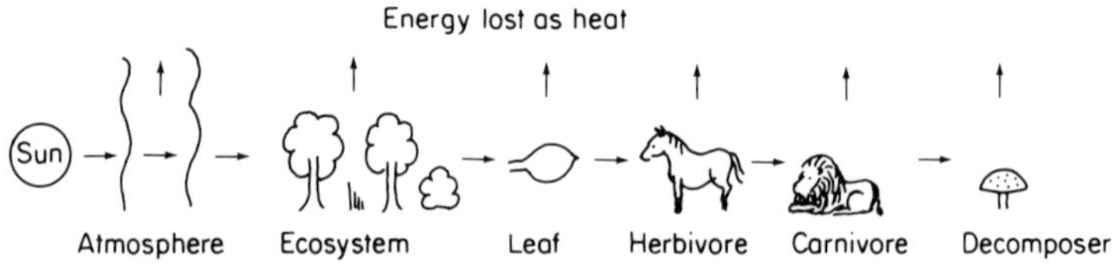

Energy lost as heat
Sun
Atmosphere
Ecosystem
Leaf
Herbivore
Carnivore
Decomposer

Fig. 4.6
The energy flow diagram showing loss of energy as heat at every stage of the chain

Photosynthesis

Photosynthesis consists of three main processes – the photochemical or light reactions, the enzymatic or dark reactions, and the exchange of carbon dioxide and oxygen between the chloroplasts and the external atmosphere. This last process, known as *diffusion*, is discussed in the following chapter. The overall process of photosynthesis can be summed up in the equation:

$$6CO_2 + 6H_2O \rightarrow C_6H_{12}O_6 + 6O_2$$

in which carbon dioxide and water are combined to give glucose and oxygen. However, due to energy considerations, the direct combination of carbon dioxide and water cannot take place in plants and the biosynthesis of glucose involves many steps, each with a specific enzyme catalyst.

The photosynthetic light reactions depend on the absorption of visible light by active pigments, of which green chlorophyll is the most important. In higher plants two photosystems operate in which chlorophyll-a molecules are in association with protein to form complexes. In photosystem 1 this complex has an absorption maximum at 700 nm, while in photosystem 2 the complex has an absorption maximum at 680 nm. By means of the electron transport reactions (details of which can be found in text books of plant biochemistry) energy rich ATP (adenosine triphosphate) is formed in which the energy from the sun is stored as chemical energy. During the light reactions water is decomposed releasing oxygen, which is passed to the external atmosphere.

The dark reactions convert carbon dioxide to energy rich sugars, a process known as *carbon fixation*. The energy required is obtained from the ATP formed during the light reactions.

The majority of plants fix carbon dioxide directly as sugars by a process known as the pentose phosphate or C_3 pathway, the first sugars formed being trioses. However, some tropical plants, notably many members of the Gramineae family, fix carbon dioxide as oxaloacetic acid, a dicarboxylic acid with four carbon atoms. This process is known as the Slack and Hatch or C_4 pathway. Plants in which the pentose phosphate pathway is the only means of fixing carbon dioxide are referred to as C_3 plants, while those which also have a Slack and Hatch pathway are known as C_4 plants.

Oxaloacetate is converted to other dicarboxylic acids which are subsequently broken down to release carbon dioxide for use in the C_3 pathway. The existence of a C_4 pathway in tropical plants means that the concentration of carbon dioxide is not a limiting factor, as it can be with C_3 plants, the gas being stored in the form of dicarboxylic acids. Thus a more efficient use of available sunlight can be made.

For most plants photosynthesis can only occur during the hours of daylight, but some tropical plants growing in arid conditions are able to fix carbon dioxide at night, although they can only obtain ATP during the hours of daylight. The fixed carbon dioxide is stored in vacuoles in the form of dicarboxylic acids until it can be converted to sugars, when sufficient ATP becomes available during the following day. Because this process was first discovered in succulent members of the Crassulaceae family, it is known as the *Crassulean Acid Metabolism* or *CAM pathway*, and species with such a pathway are referred to as CAM plants. CAM plants have an advantage over other plants growing in arid areas as they shut their stomata during the hours of daylight and thus conserve water.

Light Compensation and Light Saturation Points

In plants, energy which is stored in chemical bonds by the photosynthetic process is released during respiration. Respiration is a continuous process in which carbon compounds are oxidized to liberate energy as heat and for the maintenance of life. Whenever a plant is unable to photosynthesize, its dry weight decreases due to respiration. The amount of light required for the production of carbon compounds by photosynthesis to equal the loss of such compounds through respiration is known as the *light compensation point*. For growth and reproduction to take place photosynthesis must exceed respiration, and thus the amount of light falling on a plant must exceed the light compensation point. However, most leaves in which the C_3 pathway operates become light saturated at about 20 per cent full sunlight. The intensity at which leaves are unable to use any further radiation is known as the *light saturation point*.

During long periods of very cloudy weather photosynthesis may lag behind respiration needs to such an extent that animals depending on the plant for food starve. In the tropics, however, such conditions are rare and plants are more likely to receive sunlight which is much more intense than the light compensation point, so that the light saturation point is exceeded. Because photosynthesis is retarded and photorespiration increased at high temperatures, the reserves of carbohydrates accumulated in C_3 plants are less under such conditions than under less bright light (fig. 4.7). C_4 plants, such as sugar cane and maize, do not have a light saturation point; they are able to use all the light falling on their leaves. Such plants do not lose carbon dioxide through

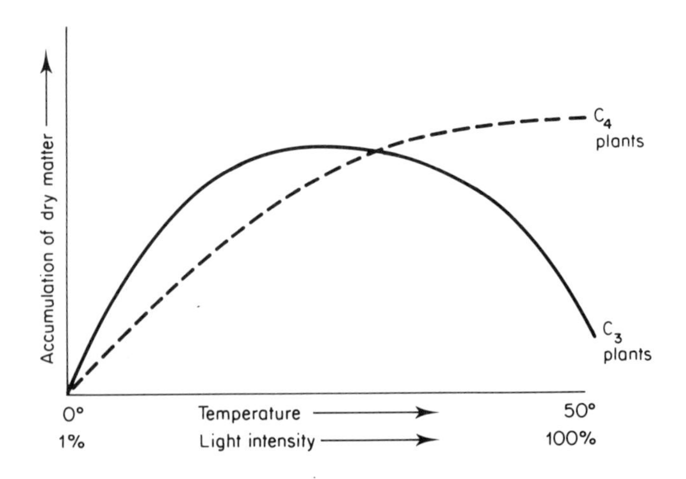

Fig. 4.7
The accumulation of dry matter by C_3 and C_4 plants at varying temperatures and light intensities

photorespiration either, as any which is released is immediately recaptured and stored in the form of dicarboxylic acids. Thus under conditions of high light intensity and temperature C_4 plants accumulate far more dry matter than C_3 plants. However, under less favourable conditions the reverse is true, as shown in fig. 4.7.

Some tropical C_3 plants, such as coffee and cocoa, produce better yields if not exposed to intense light. Thus shade trees are usually planted in coffee and cocoa plantations.

Even under conditions of bright light all the leaves of a single plant or of a plant stand will not become light saturated, as the lower leaves will always be shaded to some extent by the upper leaves. Thus when considering a complete plant or a plant stand it can be assumed that even for C_3 plants photosynthesis increases with increasing light intensity, although the upper and outer leaves may be light saturated.

Photoperiodism and Thermoperiodism

The rotation of the earth and its orbit around the sun result in cyclic changes in the amount of radiation reaching a particular place on its surface. The rotation of the earth about its axis produces alternating day and night, maximum radiation being experienced at noon, while the only radiation from the sun received by the earth at night is that reflected by the moon. Moonlight is bright enough to satisfy the requirements of some seeds for germination, to promote starch hydrolysis in leaves and to affect leaf movement in legumes.

Seasonal changes due to the orbiting of the earth around the sun are least at the equator, where plants experience 12 h of daylight throughout the year and little change in temperature, except at high altitudes. As the latitude increases, however, these changes become more pronounced, both daylength and temperature depending on the season of the year.

During their evolution, plants have come to depend on cyclic changes in daylength (*photoperiodism*) and temperature (*thermoperiodism*) to regulate some important physiological processes, including dormancy, production of new leaves in deciduous species and the onset of flowering. The alternation of day and night (*diurnal* effect) is also important to many plants.

Photoperiodism is shown by those plants whose flowering cycles depend on the daylength. At the equator all plants experience 12 h of light and 12 h of darkness, but at higher latitudes the daylengths vary throughout the year. Long-day plants require 14 h or more of daylight before they will flower, while short-day plants must have 14 h or more of uninterrupted darkness before flowering will take place. The interruption of darkness for only a few minutes is sufficient to disrupt the flowering mechanism in short-day plants. Man has made use of photoperiodism in a number of ways. Using artificial light, many plants can be brought into flower out-of-season, when they command a high price. In the U.S.A. a variety of tobacco is grown in Maryland which will not flower at that latitude, but which produces abundant, large leaves. In order to obtain seeds for further crops, plants have to be grown at the latitude of northern Florida, where they bloom abundantly but produce few leaves, these being of poor quality. Some fodder crops are grown at latitudes where they will not flower so that all the products of photosynthesis are used to produce vegetative growth. Using artificial light, plant breeders can bring plants with different day length requirements into bloom at the same time, thus enabling hybridization to take place.

In the equatorial rain forest day and night are of equal length and there is little temperature difference throughout the year. Therefore, plants have no external clock to regulate their processes. Some, such as bananas, papaya, and palms grow continuously, while amongst those that do have a period of dormancy it is not unusual to find specimens of the same species flowering and fruiting at different times of the year. On individual trees, too, branches can become dormant at different times, so that dormant shoots, new leaves, flowers, and fruits can be found on different branches at the same time.

The difference between the longest and shortest days at the tropics of Cancer and Capricorn is only 2 h, but the slight changes in daylength at latitudes only a few degrees north or south of the equator is sufficient to affect metabolic processes and development in some tropical plants. Photoperiodism, for instance, controls the level of Crassulean acid metabolism in CAM plants. For trees in the rain forests the production of new shoots and the unfolding of leaves is at a maximum twice a year at the equinoxes. The deciduous nature of the rubber tree (*Hevea brasiliensis*) is also affected by small changes in daylength, as it loses its leaves between December and March north of the equator but between June and August south of the equator. Trees growing on the equator lose their leaves intermittently.

In the tropics, where temperatures vary but little throughout the year, a sudden drop can have a surprising influence in triggering flowering activity. Coffee, certain orchids, and other plants may flower gregariously a week or so after a thunderstorm which ends a period of dry weather. It seems it is the shock of sudden coldness which initiates this reaction, rather than the increased moisture.

Differences in temperature between day and night (*diurnal periodism*) are important to both tropical and temperate plants. The kapok tree (*Ceiba pentandra*) flourishes when there is little diurnal difference in temperature. Coffee and maize, however, need night temperatures 5–10°C cooler than day temperatures to produce the best yields. Even larger temperature differences are beneficial to soybeans (*Glycine max*). In the latter case it has been shown that high day temperatures favour photosynthesis, while low night temperatures reduce respiration, a process which destroys the products of photosynthesis. Thus high yields of carbohydrates, and hence growth, are maintained when day temperatures are higher than those at night. However, this theory does not hold for all plants, as some require equal day and night temperatures, while a few, such as the African violet (*Saintpaulia ionantha*) need night temperatures higher than those of day. *Salvia splendens* requires fluctuating temperatures but it makes no difference whether it is the day or night temperature which is the lower! Despite these few exceptions, however, most plants require higher day temperatures than those at night in order to make good vegetative growth. Flowering, fruiting, and seed germination also respond to alternating temperatures.

The Effects of Altitude

Within the tropics increasing elevation reduces the duration of the daily period when temperatures rise above the threshold for plant activity. As the average temperature in mountainous regions decreases by about 0.5°C for each 100 m above sea level, even at low latitudes there is an upper altitudinal limit above which no plant will grow. This limit is set by a number of factors, including:

a. Extremes so low as to kill protoplasm.
b. The inability of the plant to accumulate compounds necessary for growth because of low temperatures.
c. Inhibition of reproduction due to low temperatures.
d. Parasites which become active only at low temperatures.
e. Adverse combination of day and night temperatures.

Even at temperatures well above freezing plants characteristic of the tropical lowlands can be injured by chilling, when membrane damage and interference with metabolism and energy transfer occurs. Within the tropics plants are unlikely to suffer from the formation of ice crystals within their tissues, except at very high altitudes. The formation of such crystals disrupts membranes and has the same effect on protoplasm as desiccation (see Chapter 3).

Plants able to withstand the effects of low temperatures normally grow at high latitudes where winter frosts occur regularly. It is not surprising, however, that genera and species normally found at such latitudes also occur in the tropics at high altitudes.

We have seen in Chapter 3 that warm air meeting a mountain mass is forced upwards and cooled, resulting in precipitation of some of the moisture carried. Thus the windward lower slopes of mountains usually have a wetter climate, and hence a lusher vegetation, than the surrounding regions. As the altitude increases the three storeyed rain forest gives way to a two storeyed submontane forest which contains fewer species of trees but an abundance of epiphytes.

The cloud belt, which occurs at high altitudes, is both the wettest region and has the least variation in daily temperature. The elfin or mist forests contain single storeyed, stunted trees festooned with epiphytes, mostly mosses, lycopods, and ferns. Where permanent cloud occurs the vegetation is continually bathed in moisture, but little radiation reaches leaf surfaces so that growth is slow.

Above the cloud belt daily temperature differences between day and night are too large for tropical lowland species to survive. Tropical montane vegetation contains many plants which also grow in temperate regions. On the mountains of Africa, which experience night temperatures of around −6°C and day temperatures of 10–12°C, a special montane moorland vegetation of *Senecio* species, *Lobelia* species, and giant heaths occurs. These plants are specially adapted to withstand frost and snow at night and intense radiation during the day. The tree senecio (*Senecio keniodendron*) for instance, has long stems protected by a covering of dead leaves. The large, woolly leaves open during the daytime to reflect radiation, while at night they close, insulating the growing shoot and maintaining the temperature of the plant above 1°C. The leaves of *Lobelia* species also protect the shoot at night, while the flowers are buried amongst bracts which have insulating properties.

At very high altitudes the increase in the intensity of light reaching plant surfaces has the effect of stunting growth, typical plants of the alpine meadows being small with their leaves growing in rosettes and all the shoots of more or less equal length. Such *cushion plants* are admirably adapted to withstand an alpine environment (Plate 4.1).

As the altitude increases, alpine meadows give way to scanty grassland and desert, which stretches as far as the snowline, above which no vegetation will grow. The changes of vegetation type with altitude are summarized in fig. 4.8.

Plate 4.1
An example of a cushion plant. (Reproduced from Plants and Environment – A textbook of Autecology by R. F. Daubenmire, 3rd edition, 1974, by permission of John Wiley & Sons Inc.)

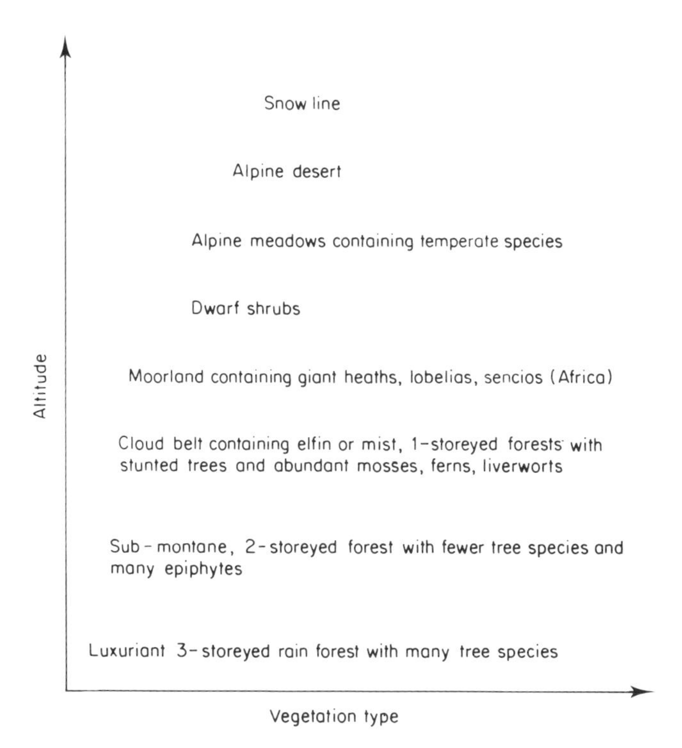

Fig. 4.8
The change of vegetation with altitude in tropical mountains

Heliophytes and Sciophytes

Plants may be classified ecologically according to their light requirements, those needing full sunlight for good growth being known as *heliophytes*, while those that grow best in shade are known as *sciophytes*. However, some plants are more adaptable than others so that there are heliophytes which will grow in partial shade and sciophytes which are not damaged by bright light. In both cases, however, the plants grow best in positions where their light requirements are most nearly met. Pioneer species of tropical forests are heliophytes, while secondary species, which colonize the area after establishment of heliophytic trees, are sciophytes.

As radiation enters a plant stand it is considerably reduced, that reaching the forest floor having an intensity of only about 1 per cent of the light experienced by the upper canopy. There are few plants which can grow under such low intensity, so that the floor of the rain forest is usually devoid of plants, except where a large tree may have fallen and allowed light to enter.

The reduction of light by a canopy of vegetation is very important ecologically, particularly after the intensity is reduced to about 20 per cent. However, because other factors such as relative humidity, soil moisture, temperature, and wind also vary within a plant stand, it is difficult to determine the influence of the light factor alone. When the term *shade* is used, therefore, it must be remembered that it includes a number of factors affecting plant growth, of which reduction in light intensity is only one.

Most of the light received by the lower storeys of vegetation within a tropical forest is the result of sunflecks and reflected radiation from leaves higher up. In general, leaves only transmit about 10 per cent of the light falling on their surfaces. Illumination in microclimates fluctuates widely. Under a canopy of vegetation the movement of leaves by wind, together with the variations in the movement of sunflecks and shadows across the ground, results in rapid and wide variations in the amount of light energy received at any given point. Thus, the light intensity at a leaf surface may rise abruptly from 2 per cent to 35 per cent for a few minutes and then drop back to 2 per cent again. Plants within a stand, therefore, must be adapted to low and/or variable light intensities, such adaptations making the most efficient use of all available radiation, as described below.

Leaves orientated horizontally produce almost complete shading of lower leaves, an effect which can be minimized by alternating the leaves on the stem (fig. 4.9). Because the lower leaves on densely leaved plants cannot photosynthesize efficiently they are often lost as the plant grows. This is particularly true of the palms, which regularly lose their lower leaves.

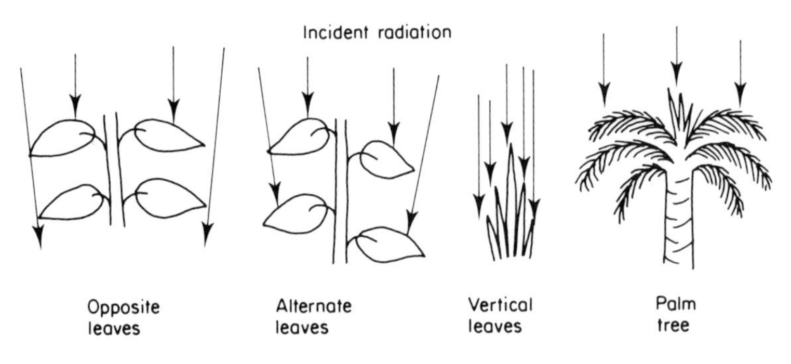

Fig. 4.9
The effects of leaf orientation on the amount of light received by lower leaves

Leaves orientated vertically allow light to enter a plant stand. Thus small herbs are able to grow amongst grasses (fig. 4.10).

Among tall plants, light requirements are critical at the seedling stage and the seedlings of most of the tallest trees of the tropical forest perish unless they germinate in a clearing receiving direct sunlight. It has been suggested that most tall forest trees became established before the climax stage.

There are many morphological and physiological differences between plants grown in full sunlight and in shade, whether the comparison is made between true heliophytes and sciophytes or plants of the same species grown under different light conditions. In comparison with plants grown in the shade, those developing in full sunlight usually show the following characteristics:

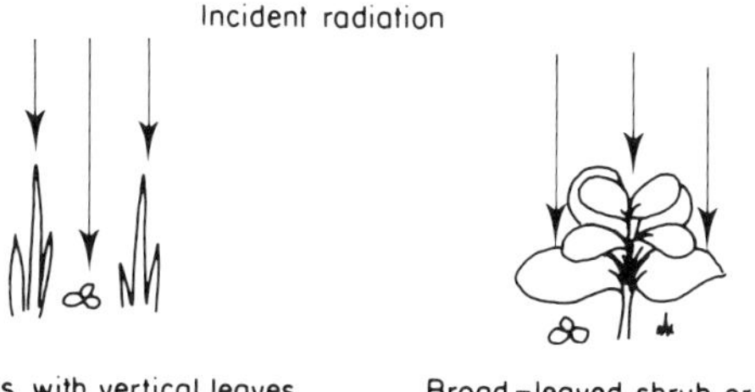

Fig. 4.10
The effect of leaf orientation on the amount of light received within a canopy

a. Thicker stems.
b. Less leaf area and shorter internodes.
c. More branching.
d. Smaller leaf blade cells with fewer and smaller chloroplasts and a greater ratio of internal to external leaf surface.
e. Thicker cuticle and cell walls.
f. Longer, more numerous and more branched roots, with a greater root/shoot ratio.
g. Leaves yellower due to lower chlorophyll content.
h. Higher photosynthetic rate per unit area of surface in bright light, but lower in weak light.
i. High respiration rate and thus higher compensation point.
j. More rapid rate of transpiration and lower percentage of water within tissues.
k. Greater vigour of flowering and fruiting.
l. Greater resistance to temperature injury, drought and parasites.

Heliophytes are capable of a more efficient use of high light intensities than sciophytes. Tropical sun-loving plants, such as sugar cane, never reach photosynthetic saturation under natural conditions, however intense the light. Sciophytes, however, often reach saturation level at light intensities of only 20 per cent full sunlight. In order to make the most efficient use of the light available, sciophytes develop large leaves with extensive surfaces containing a high concentration of chlorophyll and accessory pigments. Photosynthesis in such plants is well adapted to fluctuating light intensity, while leaf movements of some plants ensures that the best use is made of light available from sunflecks and reflection. Some herbs growing on the forest floor have permanent anthocyanin pigmentation which enables them to absorb a high proportion of the available radiation. The low respiration rates of shade plants ensures that the products of photosynthesis are conserved.

Sciophytes are often unable to survive full sunlight because their rate of chlorophyll production is too slow to balance the decomposition of the pigment by bright light.

The shading of heliophytes, such as maize, strongly reduces growth and reproduction and hence yield. In the sunflower (*Helianthus annuus*) shading reduces the rate of cell division, resulting in a smaller plant.

Although the intensity of light is the most obvious factor influencing the growth of heliophytes and sciophytes, it must also be stressed that other factors, such as water, temperature, and relative humidity vary with habitat and can be limiting. Plants growing in full sunlight must have an efficient water conduction mechanism and adequate soil moisture to survive, unless specially adapted to withstand drought. Plants growing in the shade transpire at a slower rate, which is further reduced if the relative humidity is high, as within the tropical rain forest. If sciophytes are exposed to full sunlight, however, they can quickly develop an adverse water balance, causing the stomata to close and photosynthesis to cease. There are instances when the moisture factor is critical in the shade, however, especially for seedlings of heliophytes. Such plants need to photosynthesize at a high rate in order to grow. If unable to accumulate sufficient carbon compounds these are retained by the shoots, so that the roots are starved and unable to penetrate beyond the area of surface drought.

Adaptation to High Light Intensity

As a high proportion of the light energy absorbed by plants is converted to heat, adaptations which protect the plant from high light intensities also protect against too high a temperature. The leaves of heliophytes are usually orientated so that they are not at right angles to the sun's rays, thus reducing the amount of direct radiation falling on the surface. At temperatures above about 30°C some tropical species of Mimosaceae and Caesalpiniaceae fold their leaves, thus reducing absorption of radiation even further. In leaves exposed to bright light the disc-shaped chloroplasts tend to become orientated in such a way that only one edge is exposed to direct light.

The decrease in chlorophyll content that accompanies bright light also has a beneficial effect as it results in less light being absorbed and more transmitted, thus reducing the amount of light energy which is converted to heat. It has been observed in many plants that the formation of red anthocyanins is directly correlated with light intensity. These pigments, which are located in the superficial layers of the cells, reflect red light, and since it is the long wavelengths which have the greatest heating effect, their reflection does much to protect against overheating.

Plants growing in the tropics in full sunlight are adapted in several ways to withstand the effects of high radiation. Some orientate their leaves vertically, while others have a white or shiny leaf surface which reflects much of the radiation received (Plate 4.2). A glaucous, waxy coating reflects a high proportion of ultraviolet, visible, and near infrared light. A thick coating of hairs acts in a similar manner and also shades living cells. The high transpiration rates of heliophytes ensures that the leaves are continually cooled, while the thick cuticles of some leaves and the corky bark of some trees help to insulate the plants.

Plate 4.2
Plants adapted to intense radiation. Note the white, highly reflective surfaces and the vertical orientation of the leaves. (Reproduced from Plants and Environment – A textbook of Autecology by R. F. Daubenmire, 3rd edition, 1974, by permission of John Wiley & Sons Inc.)

The Effects of High Temperature

In general, plants can be divided into three categories depending on their tolerance of heat. Heat sensitive plants are usually injured at temperatures above 30–45°C. Many lower and soft-leaved plants belong to this category. Plants growing in habitats exposed to full sunlight are heat tolerant and can survive temperatures of 60°C for short periods. However, temperatures between 60°C and 70°C are lethal for all normal plant cells with nuclei. Some bacteria and blue–green algae are heat resistant and can live at temperatures of 90°C in hot springs. These lower plants contain heat resistant nucleic acids and proteins.

Many plants are able to adapt to heat stress, so that heat tolerant species, or varieties of heat sensitive plants, have evolved in habitats exposed to high radiation.

Although, in general, adapted to the high temperatures experienced in the tropics, even plants growing naturally in such regions can suffer from high temperature injury. For example, some desert plants commonly attain leaf temperatures high enough to cause injury. In order to keep leaf temperatures below those at which injury occurs, plants such as *Citrullus* need to transpire at a high rate, and thus easily suffer from drought symptoms.

The first effect of heat stress is an increase in the reaction rates of metabolic processes or the reversal of relative rates, so that an essential metabolite is destroyed. Decreases in photosynthetic rates and the assimilation of nutrients and increases in respiration rates eventually lead to growth reduction and starvation. Such effects, due to moderately high temperatures occurring for hours or days, are generally reversible.

Very high temperatures experienced for only a few seconds can cause heat shock. The main effect of heat shock is to damage membranes by the breakdown of protein and the increased mobility of lipids. Such effects are not reversible, the organ affected, or even the whole plant, eventually dying.

Transpiration

Transpiration by plants is influenced by radiation both in the form of heat and as light. The rate of transpiration depends on the difference between the temperature of the leaf surface and that of the surrounding air. The higher the temperature of the leaf the greater will be the amount of water lost. Temperature also changes the ratio of cuticular to stomatal transpiration: the higher the temperature the greater the cuticular component. Thus at a temperature of 49°C the nocturnal rate of transpiration in the sunflower reached 91 per cent of the daytime rate, even though the stomata were closed.

High light intensities also promote rapid transpiration, as light stimulates the opening of the guard cells and increases the permeability of the plasma membranes. Of the light falling on a leaf surface about two-thirds is converted into heat which is reradiated or used to warm the leaf surface and vapourize water. Thus it is impossible to separate the effects of heat and light, but an increase in total radiation will always increase the rate of transpiration, as long as the water balance of the plant does not fall to a dangerously low level. Should this occur the stomata will close (see Chapter 3).

The Effects of Low Temperatures

The ability of plants to endure low temperatures varies widely among species. Tropical plants such as rice, cotton, etc. are injured by exposure to temperatures which, although low, are well above freezing. Other plants commonly found in temperate regions are not injured until they are frozen, while those native to cold regions can survive periods when their tissues are frozen solidly. There is no region on earth which is either too hot or too cold for some plant to grow, though these will be lower plants, higher plants being unable to survive in extreme habitats.

When the temperature drops below the minimum for growth a plant becomes dormant, even though respiration and sometimes photosynthesis continue slowly. At such temperatures chlorosis due to cold injury is likely to occur. With a further decrease in temperature a point is reached when the protoplasm is killed. The main cause of fatal cold injury in tropical plants is the precipitation of proteins.

Many plants become susceptible to disease only above or below certain temperatures. Maize, for example, becomes very susceptible at temperatures below 13°C.

Reproduction and Germination

Many temperate plants will not reproduce in tropical climates, even if they are able to grow at such temperatures, because they need a period of cold to stimulate the formation of flower buds. For example, *Calceolaria* and *Senecio cineraria* will not form flower buds if the temperature is kept above about 15°C.

In the tropics where temperatures vary little throughout the year, small deviations are effective in initiating flowering, as has been described above. Light also influences flowering, some plants only flowering if exposed to bright sunlight. In climates with a high percentage of cloudy days such plants remain vegetative. For others, the number of hours of darkness is critical, short-day plants requiring at least 14 h of darkness and long-day plants less than 10 h of darkness.

The seeds of many plants of cold regions require chilling under moist conditions for a period before they will germinate, while those of most plants become sensitive to light when wetted. In some instances this is beneficial, while in others it is not. Jute (*Corchorus olitorius*) and lettuce will not germinte unless exposed to light, but the vanilla orchid (*Vanilla fragrans*) and many members of the Liliaceae family must have total darkness. For some plants the exposure to light need be very short indeed – only 0.01 s allows some tobacco seeds to germinate.

However, although germination responses to light are controlled by a pigment sensitive to red light, so many aspects of the environment influence this vital process that it is difficult to pinpoint the effects of light alone with any certainty.

Phototropism

The absorption of blue and ultraviolet light rays by blue light receptors, such as the carotenoids and riboflavin, is responsible for the phenomenon of *phototropism*. Plants showing phototropism turn towards the direction of greatest, constant light intensity, as shown in fig. 4.11. The absorption of blue and ultraviolet light depresses growth, this being greatest on the side of the shoot facing the light, so that the shoot bends.

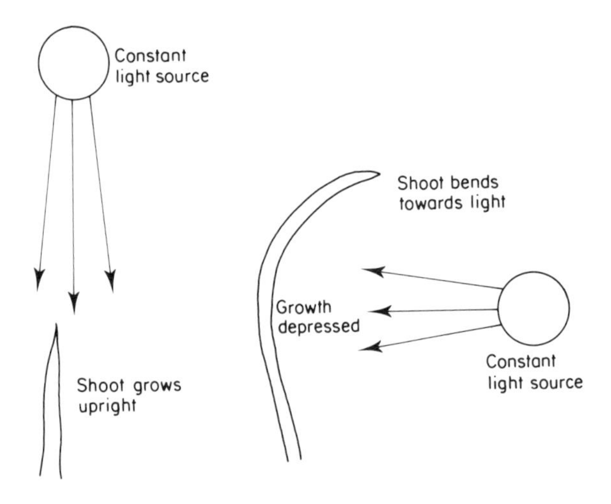

Fig. 4.11
Phototropism

Suggestions for Further Reading

Burris, R. H. and Black, C. C. (Eds) (1976). *CO_2 Metabolism and Plant Productivity*. University Park Press.

Etherington, J. R. (1975). *Environment and Plant Ecology*. Wiley.

Etherington, J. R. (1978). *Plant Physiological Ecology*. Arnold.

Hall, D. O. and Rao, K. K. (1981). *Photosynthesis* (third edition). Arnold.

Larcher, W. (1980). *Physiological Plant Ecology*. Springer-Verlag.

Levitt, J. (1980). *Responses of Plants to Environmental Stresses*, Vols. 1 and 2. Academic Press.

Longman, K. A. and Jenik, J. (1974). *Tropical Forest and its Environment*. Longman.

Treshow, M. (1970). *Environment and Plant Response*. McGraw-Hill.

Turner, N. C. and Kramer, P. J. (1980). *Adaptation of Plants to Water and High Temperature Stress*. Wiley.

Sutcliffe, J. (1977). *Plants and Temperature*. Arnold.

Villiers, T. A. (1975). *Dormancy and the Survival of Plants*. Arnold.

Whatley, J. M. and Whatley, F. R. (1980). *Light and Plant Life*. Arnold.

Chapter 5

Plants and the Atmosphere

Introduction

The unique mantle, known as the *atmosphere*, which surrounds the earth's surface contains a mixture of gases, collectively known as *air*, together with dust particles and water vapour. Without the atmosphere, life, as we know it, could not exist on earth. The molecules of gases and particles of dust act as a protective blanket, preventing lethal radiation reaching the earth's surface and reducing the large diurnal fluctuations in temperature which would otherwise occur. Such fluctuations would be too great to permit the existence of any known form of life. The atmosphere is also the source of water for all land plants and animals, and of oxygen, needed by most plants and animals for the maintenance of life.

At its birth the composition of the earth's atmosphere was totally different from that at the present time. As the earth's surface cooled, however, the early gases were gradually replaced by water vapour, carbon dioxide, and nitrogen. Most of the water vapour condensed to form the oceans and the carbon dioxide became fixed in carbonate minerals. Oxygen did not appear in the atmosphere until well after the advent of the first photosynthesizing plants. That present initially was quickly removed as oxide minerals, but eventually the earth's surface became saturated and free oxygen could exist in the atmosphere, thus enabling the evolution of higher plants and animals.

Today the air of the atmosphere contains approximately 79 per cent nitrogen, 21 per cent oxygen, and 0.03 per cent carbon dioxide. The proportions of these gases remain fairly constant, but other constituents of the atmosphere, such as water vapour, dust particles, volatile substances, and pollutants can vary greatly.

The constancy of the ratio of oxygen to carbon dioxide in the atmosphere indicates that a global equilibrium exists, which, so far, has not been upset by the activities of man, despite the increase in the burning of fossil fuels. The burning of fossil fuels and organic matter absorbs oxygen and releases carbon dioxide. It is probably the buffering action of the oceans with their enormous capacity for algal photosynthesis and their absorption of carbon dioxide as carbonates, which maintains this gaseous equilibrium.

The Importance of Atmospheric Gases to Plants

The presence of carbon dioxide in the atmosphere is of paramount importance to plants and indirectly to animals, as without this gas there would be no synthesis of carbohydrates and, therefore, no storage of chemical energy to maintain life. Oxygen is also of importance to most plants and animals, as it is only through the oxidation of carbohydrates during the process of respiration that the stored chemical energy can be released. However, there are some lower plants, the anaerobic organisms, which do not need atmospheric oxygen for respiration. Some, such as the denitrifying bacteria, obtain their respiratory oxygen from nitrates, while others are able to dispense with oxygen altogether and use hydrogen sulphide. Such organisms make up only a very small proportion of the total plant population, however, and the majority of plants must receive adequate supplies of oxygen through both aerial parts and roots.

To most plants and animals the nitrogen content of the atmosphere is only important in so far as it dilutes the other gases to acceptable concentrations. However, some lower plants, the nitrogen-fixing bacteria, and algae, are able to convert gaseous, atmospheric nitrogen into forms which can be used to build proteins, essential constituents of life. Some of these, such as the *Azobacter* group, which contains the tropical *Beijerinckia* bacteria, are free living in the soil. Most free-living nitrogen-fixing bacteria are not, however, of great environmental importance, as their contribution to soil nitrogen is small. The symbiotic nitrogen-fixing systems described in Chapter 7, however, are of considerable ecological importance. For example, colonies of nitrogen-fixing bacteria live in the root nodules of leguminous plants, and their hosts are able to use the resulting nitrogen compounds to synthesize proteins.

The Oxygen–Carbon Dioxide Cycles

Oxygen is absorbed during respiration by plants and animals and the burning of any substance, while it is released during photosynthesis. Carbon dioxide is absorbed during photosynthesis and released during respiration, burning of organic matter, and the decomposition of organic matter and carbonate rocks. The oxygen–carbon dioxide cycles are summarized in fig. 5.1, which shows the extreme importance of photosynthesis in maintaining the concentration of atmospheric oxygen, which would otherwise be soon depleted.

Carbon Dioxide and Photosynthesis

The amount of carbon dioxide in the normal atmosphere is lower than that which could be used by plants under optimum conditions. Within a forest ecosystem, however, especially the tropical rain forest, the concentration of carbon dioxide is several times greater than normal. Such an increase probably compensates for the reduced light intensity within such a forest and ensures that shade plants can utilize the light they do receive to the maximum extent.

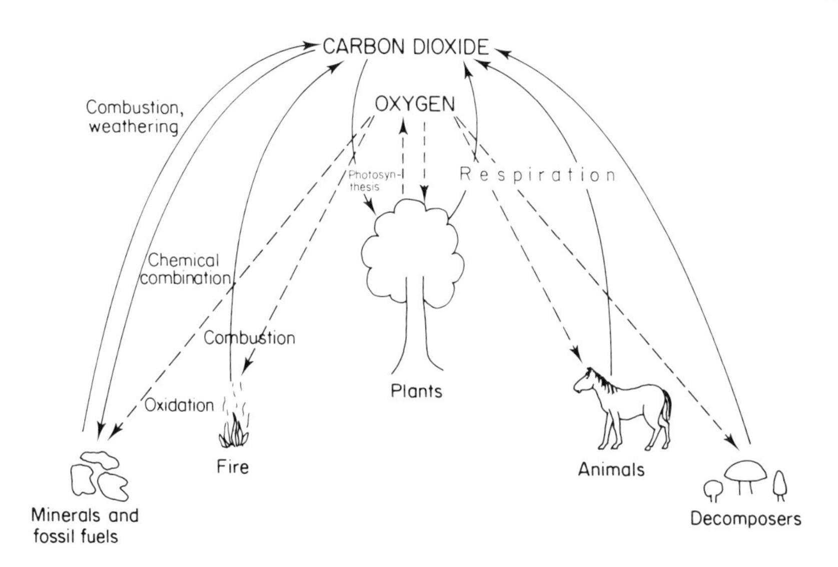

Fig. 5.1
The oxygen–carbon dioxide cycles

Within a plant carbon dioxide is absorbed during photosynthesis and released during respiration. Unless the sky is very overcast or the plant growing in dense shade, the amount of carbon dioxide absorbed during the hours of daylight exceeds that evolved during respiration. At night, however, only respiration takes place in most plants, so that carbon dioxide is evolved but none is absorbed. The point at which photosynthetic absorption of carbon dioxide equals respiratory evolution of the gas is known as the *compensation point*.

During photosynthesis oxygen is evolved, while it is absorbed by the respiratory process. Thus two complementary processes take place in plants involving the exchange of gases with the atmosphere (fig. 5.2). These exchanges take place by the process of diffusion.

During the hours of daylight an enormous volume of air is required by plants, as for every gramme of glucose formed the carbon dioxide contained in 2500 l of air must be absorbed. Thus an inadequate supply of carbon dioxide is often the limiting factor for photosynthesis, especially in the tropics.

The amount of oxygen in the air is always sufficient for the respiration of aerial parts of plants, although it can be inadequate for roots, due to waterlogging or compaction of soil.

Plants can be classified as C_3 or C_4 according to the mechanisms they use to convert carbon dioxide and water into sugars (see Chapter 4). Most are C_3 plants. Such plants use only the pentose phosphate pathway, the first organic compounds formed from carbon dioxide and water being trioses with three carbon atoms. However, some plants, including many tropical savanna grasses, fix carbon dioxide as oxaloacetate, a compound with four carbon atoms. These

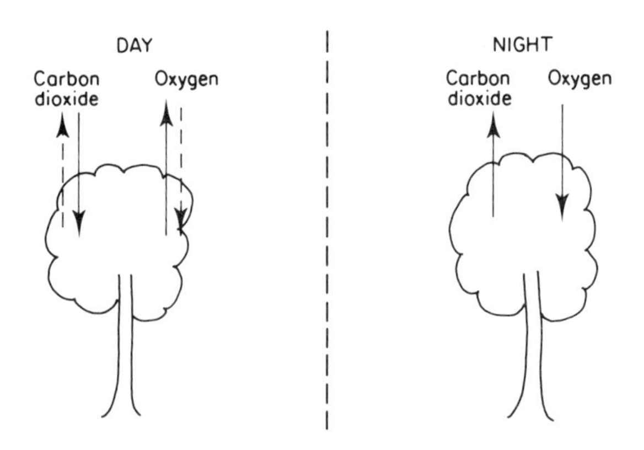

Fig. 5.2
Gaseous exchange in plants

are known as C_4 plants. Some succulents, especially those belonging to the Crassulaceae, are able to fix carbon dioxide at night and store the gas in the form of organic acids. These acids are subsequently broken down and the released carbon dioxide used in the pentose phosphate pathway. Such succulents are known as *CAM plants* (Crassulean Acid Metabolism).

Because C_3 plants are unable to store carbon dioxide they must have a regular supply for efficient photosynthesis. Thus the absorption of the gas becomes the limiting factor in tropical savanna habitats, where sunlight is abundant. Due to increased rates of photorespiration C_3 plants also rapidly lose the products of photosynthesis when exposed to high light intensities.

In contrast to C_3 plants, C_4 plants can store carbon dioxide and are thus able to maintain high rates of photosynthesis at high temperatures and light intensities. Also, the products of photosynthesis are not lost through photorespiration. Thus C_4 plants are admirably adapted to the savanna and semi-desert environments of the tropics.

CAM plants close their stomata during the day, reducing transpiration considerably without at the same time stopping the photosynthetic process, as carbon dioxide is absorbed and stored at night. This is an important adaptation for plants of desert and semi-desert habitats.

Plant Stomata

The exchange of gases between plant cells and the atmosphere can only take place through openings in the leaf surfaces known as *stomata*. Stomata are thus the most important regulators of the diffusion process. The number, size and distribution of the stomata on a leaf varies not only with the species but also with habitat, so that individuals of the same species growing in different habitats can have different stomata patterns.

As carbon dioxide enters a leaf through the stomata it dissolves in the cell sap and slowly migrates to the chloroplasts, where photosynthesis takes place. In still air a boundary layer of air exists over the surface of a leaf. This can become almost devoid of carbon dioxide if the gas is able to diffuse through the stomata more quickly than it is replaced by diffusion from the outer air (fig. 5.3). The thickness of the boundary layer depends on air movement, the more windy it is the thinner the layer will be. As carbon dioxide can only enter a leaf through the stomata no photosynthesis can take place when the stomata are closed, except in CAM plants.

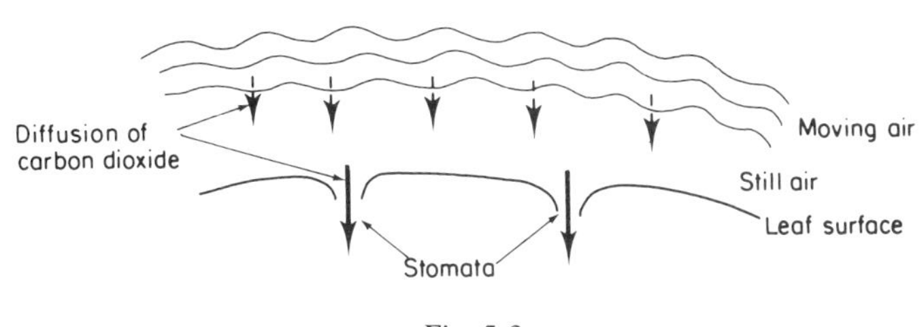

Fig. 5.3
The diffusion of carbon dioxide

Carbon dioxide produced during respiration can be under such a high pressure that it is forced out through the cuticle, especially at night when the stomata of most plants are closed. As photosynthesis and respiration take place in different parts of a cell the carbon dioxide produced by respiration cannot be used directly for photosynthesis.

The rate at which gases diffuse into and out of a leaf depends on the extent to which the stomata can open, the maximum width being known as the *pore width*. The pore width is large in the leaves of tropical forest trees and small in sclerophyllous plants. As the number of stomata per unit area is also high in tropical forest trees the total area through which gases can diffuse can be as much as 3 per cent of the leaf surface. For most plants, however, the pore area is only around 1 per cent, while for succulents with few stomata it can be 0.5 per cent or less.

The opening and closing of the stomata depend on two main processes: the water balance of the plant and the partial pressure of carbon dioxide within the intercellular space. During the hours of daylight most plants use up carbon dioxide, so that the partial pressure within the intercellular space falls, causing the stomata to open. At night the reverse is true and the stomata close. In CAM plants carbon dioxide is used up at night so the stomata open. During the day this gas is released from stored organic acids, causing the partial pressure within the intercellular space to rise and the stomata to close.

However, despite the regulating power of carbon dioxide the most important influence on the opening of the stomata is the water balance of the plant (see Chapter 3). If this is negative the stomata will not open whatever the partial pressure of carbon dioxide. Other factors also influence stomatal opening and the interactions between all factors are such that conditions are seldom right for maximum pore width to be achieved. In fact, plants growing in extreme habitats, such as deserts and high altitudes may be so adversely affected that their stomata remain closed for long periods.

Primary Productivity

Plants consist of 60 per cent or more of carbohydrates, which are the products of carbon dioxide fixation through photosynthesis. All animals obtain their carbohydrates either directly or indirectly from plants. Thus plants are the primary producers and the quantities of carbohydrates they accumulate are of paramount importance to animals, including man. The rate at which plants accumulate dry matter is termed the net primary productivity (see Chapter 1) and is measured as grammes of organic dry matter per square metre of ground. Organic dry matter contains all the organic compounds synthesized by plants and includes proteins, fats, and secondary metabolites as well as carbohydrates. However, not only do carbohydrates make up the bulk of dry matter but all other compounds are ultimately derived from sugars. At high temperatures and light intensity the primary productivity of tropical grasses with a C_4 pathway is greater than that of C_3 plants growing in the same environment, while CAM plants, such as the pineapple, have characteristically low net primary productivities.

In order for a plant to build up large carbohydrate reserves the time over which a high rate of carbon dioxide absorption can take place is important. This is affected by many factors, including hours of darkness, cold and drought, as well as loss of leaves in deciduous species. Evergreen trees of the tropical forest are least affected by such adverse conditions and are thus able to build up high reserves of carbohydrates and other metabolites. Shade plants and plants of arid, montane, or arctic regions, however, are greatly affected by adverse conditions for long periods. Thus their production of dry matter is small and net primary productivity can even be negative, if photosynthesis is reduced to such a level that the rate of dry matter production is less than its destruction through respiration. A plant maintaining a high proportion of non-green organs, such as flowers, fruits, woody stems, and roots will also lose much of its photosynthetic production, as such organs use up carbohydrates during respiration but are unable to produce these compounds by photosynthesis.

The Soil Atmosphere

All normal soils contain spaces between particles which are filled with a mixture of gases known as the soil atmosphere. However, the proportions of the various gases are not the same in the soil atmosphere as they are above ground. The soil atmosphere is usually saturated with water vapour, while the respiration of soil organisms and plant roots increases the proportion of carbon dioxide, which can be as high as 13 per cent. As no photosynthesis can take place below the surface of the soil, oxygen is used up during respiration but is not replaced. Thus the proportion of this gas in the soil atmosphere is generally lower than above ground.

The diffusion of gases through the pore spaces of a soil is slow, so that, although gas exchange is continuously taking place with the outside air, the carbon dioxide content of most soils remains high, while that of oxygen is usually low. The rate at which soil gases diffuse and are exchanged depends on soil structure (see Chapter 2). In moderately coarse or well aggregated soils the large pore spaces enable a more rapid diffusion of gases than in the small pore spaces of a poorly aggregated soil. In the latter, the concentration of carbon dioxide may reach toxic levels, while the deficiency of oxygen can be limiting for many plants and aerobic soil organisms. Although fine textured soils have more pore space than coarse soils, the smaller pore size retards diffusion considerably. Large pores allow water to drain downwards quickly, but small, capillary pores hold water against gravitational forces. The numbers of large pores represent the minimum air capacity of a soil and are thus of extreme importance to the aeration of a soil. Regardless of the number and size of the pores a soil will be poorly aerated if drainage is obstructed and the water table is near the surface.

In the tropics, the diffusion of gases can be so rapid in sandy soils with large pore spaces that a higher than normal concentration of oxygen is maintained. This allows a rapid breakdown of organic matter so that such soils are characteristically low in humus content.

When water enters a soil it fills the pores, replacing the soil atmosphere. If the water drains away quickly, air from the outside is drawn in, thus temporarily raising the oxygen level of the soil atmosphere. In heavy soils, however, drainage is so slow that pores remain filled with water for some time, reducing the space available to the soil atmosphere.

The Effects of Low Oxygen Concentration

Although poor soil aeration affects plants in a number of ways, the most important is lack of oxygen. Most soils are deficient in oxygen and some micro-organisms, such as *Clostridium*, have become so adapted to low oxygen levels that they find environments with appreciable quantities of the gas unfavourable and become inactive. Most higher plants however require adequate supplies of oxygen for their roots if they are to grow normally. The problems associated with oxygen deficiency in some soil types are magnified in the tropics, as high temperatures increase respiration rates. Respiration rates are also increased where there is abundant litter and/or humus for decomposition by micro-organisms, while roots increase their respiration rates during the growing season.

Roots cannot function if the oxygen concentration in the soil atmosphere is less than 10 per cent. In most well drained soils the concentration usually lies between 10 per cent and the 21 per cent of the outside atmosphere, with the higher values occurring near the surface. As a plant grows it sends out roots in all directions, so some at least will encounter large pores with a good oxygen supply. Rootlets develop in abundance where conditions are most favourable for the plant as a whole. A sudden change in these conditions can have a disastrous effect on the plant. If the oxygen content of the soil atmosphere is suddenly reduced the plant wilts, chlorophyll ceases to be synthesized, and eventually the plant dies. A gradual change, however, can be tolerated by most plants as the roots have time to discover more favourable areas of soil.

Below an oxygen concentration of 10 per cent the normal functions of the roots are adversely affected, while at less than 2 per cent the roots die. The oxygen content of the soil layer just above the water table is only around 1 per cent, so that roots of normal plants cannot exist in this layer. However, there are plants, such as rice and mangroves, which have become adapted to such low oxygen levels and whose roots can spread into and below the water table. Other plants may avoid the deficiency in oxygen by developing shallow root systems, but there are many which cannot grow in soils with high water tables.

Plant Adaptation to Low Oxygen Levels

Plants adapt to low oxygen levels in the soil in several ways, which include shallow rooting systems, special aerating tissues and organs, low oxygen requirements, and the ability to respire anaerobically. Most higher plants possess a continuous intercellular air space system which is linked to the outside atmosphere through the stomata. In hydrophytes this internal air system is highly developed (see Chapter 3), while some mesophytes are able to increase their air spaces when growing in water or wet soil.

The efficiency of oxygen conduction from the shoots is well illustrated by paddy rice, in which the roots may contain as much as 18 per cent oxygen while the surrounding mud contains none. Some hydrophytes, such as the black mangrove (*Avicennia nitida*) produce special root branches which grow erect until they project into the air above the mud and high water level (Plate 5.1). These structures, called *pneumatophores*, have a well developed intercellular system of air spaces which are continuous with the stomata, thus allowing gas exchange to take place. Other hydrophytes, such as the mangrove (*Rhizophora* spp.), have prop roots above the mud level which are well supplied with openings to the outside. These openings, known as *lenticels*, allow gas exchange, oxygen diffusing downwards into the submerged roots.

The ability to respire anaerobically for a short time is possessed to a limited extent by the mature tissues of most plants. It is however especially well developed in some hydrophytes which grow in still water or very wet soil. Anaerobic respiration begins when the oxygen content of the intercellular spaces drops to about 3 per cent.

Germination

Most seeds require abundant oxygen for germination. When this gas is at low concentration respiration proceeds very slowly and dormancy is prolonged.

Plate 5.1
The pneumatophores of the mangrove *Avicennia nitida* together with a small *Rhizophora mangle* tree. (Reproduced from Plants and Environment – A textbook of Autecology by R. F. Daubenmire, 3rd edition, 1974, by permission of John Wiley & Sons Inc.)

Buried seeds can remain viable but without germinating for many years, but germinate promptly when brought to the surface. Lotus seeds buried in a bog have been known to germinate after a dormancy of 1000 years.

As many seeds also require light to germinate (see Chapter 4), both lack of light and lack of oxygen can contribute to the dormancy of deeply buried seeds.

Many plants that grow best in muddy soils where the oxygen concentration is extremely low have developed especially low oxygen requirements for germination. The oxygen requirement for the germination of rice, for example, is only one-fifth of that required by wheat. This adaptation to low oxygen requirements includes the ability to respire anaerobically, a property shared by some upland members of the Leguminosae family, whose seed coats are relatively impermeable to oxygen. In the latter case only after germination has progressed far enough to rupture the seed coat can normal, aerobic respiration of the embryo begin.

Aquatic Ecosystems

The photosynthetic activity of green hydrophytes can saturate the water within their immediate vicinity with dissolved oxygen. In general, however, the water at the surface of a lake contains less than 1 per cent dissolved oxygen, while diffusion from the atmosphere to the lake bottom is a very slow process. Convection currents help to distribute the available oxygen more evenly, but generally a deficiency of oxygen exists, especially in the bottom layers of water.

Wind

The formation of winds is due to the uneven heating of land and water and the large temperature differences between the equator and the poles. Wind velocity depends on many factors, including topography, vegetation masses, position of seashores, height above sea level, and major wind paths and regions of calm.

Wind is an ecological factor of considerable importance, especially on flat plains, along sea coasts, and at high altitudes. It affects plants directly by warming or cooling the leaves, increasing or decreasing transpiration, causing various types of damage, and by scattering pollen, fruits, and seeds. Less direct effects are numerous, including the transportation of hot and cold air masses, clouds, and fog, and the modification of temperature.

The influence of wind on plants

Wind velocity affects all diffusion processes between a plant and the atmosphere. Thus the rate of exchange of gases, water, and heat depend on wind speed. The boundary layer of still air which covers the surface of a leaf at a thickness of a few millimetres impedes diffusion, but vigorous air movement thins this layer. Thus an increase in wind velocity aids the exchange of gases and, in general, increases transpiration and heat loss. However, the influences of wind are complex, as its cooling effect can also reduce water loss, thus slowing transpiration, while the reduction in humidity at the leaf surface can result in closure of the stomata. In the latter case gas exchange is slowed or ceases altogether, which results in a decreased rate of photosynthesis.

At high wind speeds with the stomata closed, cuticular transpiration becomes important and plants with thin cuticles can be desiccated. Thus plants subjected to continuous or periodic winds must develop thick cuticles if they are to survive. It is thought that the semi-desert, xeromorphic vegetation found on exposed hills in parts of West Africa is the result of the annual influence of the hot, dry wind, known as the Harmattan.

In the tropics, hot, dry winds can cause such extreme desiccation that leaves, shoots, and fruits are killed. The taller the plant, the more it is subjected to such effects. Low growing plants escape damage, cushion plants, in which all the shoots are short and of equal length, being particularly well adapted to withstand the effects of such winds.

Many of the effects of wind, such as increased transpiration, reduced photosynthesis, and increased respiration due to bending and rubbing, reduce growth and lead to a dwarf form of vegetation. Plants developing under the influence of drying winds never attain a degree of hydration that enables them to expand their maturing cells to normal size. As a result all organs become dwarfed, without necessarily becoming deformed.

Wind damage and deformation

The effect of wind blowing from a constant direction is to alter the natural shape of a plant (Plate 5.2). Tree trunks may become bent and crowns asymmetrical. Deformation is not necessarily accompanied by dwarfing, as moist winds can mould the form of a shoot without appreciably reducing its size.

Plate 5.2
The effects of wind on the shape of trees. (Reproduced from Plants and Environment – A textbook of Autecology by R. F. Daubenmire, 3rd edition, 1974, by permission of John Wiley & Sons Inc.)

In the tropics wind damage of plants can be severe, as tropical cyclones and hurricanes destroy much vegetation, while squalls that develop before thunderstorms damage plants, especially trees. The frequent hurricanes experienced by the West Indies have produced a sub-climax vegetation on the exposed slopes of mountains. Such vegetation is characterized by the absence of large trees, while on the windward sides of mountain peaks and ridges the vegetation is stunted.

The most severe form of damage is the uprooting of plants or the breakage of trunks and stems such that the plant dies. Less severe damage includes breakage of shoots, twigs, etc., defoliation, and damage to leaves, all of which lead to a decreased rate of photosynthesis and increased respiration. The large leaves of the banana (*Musa*) are particularly susceptible to damage through shredding or tearing between the parallel veins (Plate 5.3). However, such damage does not always decrease yield and can be beneficial in cooling the leaves. Under bright sunlight entire leaves can reach near lethal temperatures which severely reduce photosynthesis.

Lodging

Strong winds can also damage smaller plants, especially those belonging to the grass (Gramineae) family. Wind frequently flattens these plants against the ground, a form of damage known as *lodging*. If the stems are not too mature,

Plate 5.3
Wind damage to banana leaves. (Reproduced from Plants and Environment – A textbook of Autecology by R. F. Daubenmire, 3rd edition, 1974, by permission of John Wiley & Sons Inc.)

the prostrated plants become partially erect again by growth from the lower nodes. However, such damage can reduce yields and is thus important to crop plants, such as maize and sugar cane.

Abrasion

Winds carrying particles of sand cause abrasive damage by eroding away bark and buds. This action is strongest a few centrimetres above ground level, and crops grown on sandy soils in windy places are often damaged in this way.

Salt spray

Along sea coasts salt spray is carried ashore by wind and during severe storms it can penetrate inland for many kilometres. The worst damage to plants follows storms which are not accompanied by rain, so that a film of salt is deposited on plant surfaces. Some plants, such as the coconut (*Cocus nucifera*), have become adapted to such conditions, but others are very sensitive to the effects of salt and cannot grow near sea coasts.

Erosion

An undisturbed plant cover is very effective in preventing the movement of soil by wind. However, the effect of wind on eroding bare patches of soil can be so severe as to expose the roots of living plants. These then die, increasing the area susceptible to erosion. The eroded soil is also a hazard to the existence of plants where it is deposited, for few species can tolerate the sharp reduction in aeration about their roots that follows the deposition of new soil upon old. However, there are plants which have adapted to such a habitat. These produce adventitious roots at successively higher levels on the stem as deposition takes place (Plate 5.4).

Windbreaks

The velocity of a wind is strongly reduced by the presence of plant cover, even herbaceous plants having an effect at ground level. The effects of a forest on wind velocity can extend 100 m to the leeward side, thus projecting the forest microclimate beyond its edge. Conversely, on the windward side the microclimate within the forest is influenced by breezes entering the tree stand.

Special plantations of trees, known as *windbreaks* or *shelter belts*, are made throughout the world to protect crops, livestock, and buildings from the effects of strong winds. The effectiveness of such plantations depends on the denseness of vegetation. If too open it will have little effect, if too dense undesirable turbulence results.

Plate 5.4

An example of a tree able to withstand the deposition of soil. Note the adventitious root formed when the lower stem was covered. (Reproduced from Plants and Environment – A textbook of Autecology by R. F. Daubenmire, 3rd edition, 1974, by permission of John Wiley & Sons Inc.)

Windbreaks confer several types of benefit on the crops they protect. They reduce evaporation, transpiration, and damage to crops by breakage and lodging. By reducing the velocity of the wind, shelter belts help to reduce erosion through movement of the soil by wind.

However, windbreaks can also have adverse effects, as in their immediate vicinity they exhaust the soil of moisture and nutrients, thus reducing the amount of land available for crop planting.

Wind pollination

The most primitive pollen-producing plants are believed by some ecologists to have depended on wind to transfer their pollen from anther to stigma. Many plants are still pollinated by wind, especially those belonging to the Coniferae and Gramineae families. However, although air currents are almost always present to some extent and pollen grains can be carried for many kilometres, there are several disadvantages to wind pollination. Because of scattering by wind the chances of a pollen grain alighting on a mature stigma of the same species are very remote. Thus extremely large quantities of pollen must be produced to ensure adequate seed formation, which is wasteful of plant resources.

Wind pollinated plants have developed certain morphological adaptations which help to overcome the disadvantages of this type of pollination. The flowers have long stamens projecting beyond the perianth, so that the pollen is easily blown off by the lightest breeze. Compared with insect pollinated plants the perianth is reduced, absent, or deciduous. Colour and scent are not important and are usually absent, perianths and bracts generally being green, brown, or dark red. The stigmas are well exposed and often feathery, so that they comb the air for any pollen grains which may be present. The flowers are typically unisexual and are usually positioned high up so that they are never sheltered from the wind by foliage. Unisexual flowers prevent self-fertilization, which can occur in hermaphrodite flowers, such as those of the kapok tree (*Ceiba pentandra*). The pollen of wind pollinated flowers is very buoyant and not sticky as is that of insect pollinated plants. Some plants have developed mechanisms to help disperse their pollen. The stamens of grasses, for example, are continually moving, while the anthers of most plants only open when the weather is warm and dry, thus preventing rain from washing out the pollen. The anthers of the castor oil plant (*Ricinus communis*) are explosive, shooting pollen into the air.

Apart from the gymnosperms and the Gramineae family, wind pollination is not important amongst tropical plants. In tropical forests there is little wind, while animal pollinators abound. Whereas temperate species of *Quercus* and *Castanea* are wind pollinated, for instance, tropical species are insect pollinated.

Wind dissemination

Wind is the most efficient agent of dissemination and many terrestrial plants depend on it to scatter their seeds. Three common types of adaptation are found which help dispersal by wind. The spores of lower plants are very small and can thus be blown considerable distances. Similarly, the tiny seeds of members of the Orchidaceae and Ericaceae families are also easily blown away by wind. The fruits or seeds of many plants belonging to the Bombacaceae, Malvaceae, Compositae, and Asclepiadaceae families are covered in hairs that greatly increase their buoyancy without increasing their weight. Fruits or seeds of some trees are winged, for example, members of the Bignoniaceae family, which slows the speed of descent and enables the seeds to be carried horizontally for some distance.

Wind disseminated tree seeds occur in almost all common tropical families, such as Apocynaceae, Bombacaceae, Bignoniaceae, Dipterocarpaceae, Leguminosae, Sapindaceae, Sterculiaceae, etc. Trees within the tropical evergreen rain forest are sheltered from wind and thus their seeds are dispersed by other methods. Some top storey trees, however, which became established before the climax was attained, have wind dispersed seeds. Some vines, particularly those belonging to the Asclepiadaceae and Bignoniaceae also have seeds disseminated by wind.

Many trees of the tropical deciduous forest have wind dispersed seeds, and in contrast to the evergreen forests some shrubs and herbs beneath the canopy have seeds which are disseminated in this way, as wind is able to penetrate the canopy when the trees are leafless.

Savanna regions often experience strong winds and thus many plants characteristic of such regions have seeds dispersed by wind.

Suggestions for Further Reading

Burris, P. H. and Black, C. C. (Eds) (1976). *CO_2 Metabolism and Plant Productivity*. University Park Press.

Etherington, J. R. (1978). *Plant Physiological Ecology*. Arnold.

Faegri, K. and von der Pijl, L. (1966). *The Principles of Pollination Ecology*. Pergamon Press.

Grace, J. (1977). *Plant Response to Wind*. Academic Press.

Opik, H. (1980). *The Respiration of Higher Plants*. Arnold.

Tivy, J. (1971). *Biogeography*. Oliver and Boyd.

Chapter 6

Tropical Vegetation

Introduction

The area lying between the tropics of Cancer and Capricorn contains a wide variety of vegetation types, ranging from luxuriant rain forest to desert (see fig. 1.3). Primarily, the type of vegetation found in any tropical area depends on precipitation, both total annual rainfall and its distribution throughout the year. These factors are further modified by altitude, the vegetation found at moderate and high altitudes being quite different from that of the tropical lowlands. Soil type and wind also influence vegetation patterns, but unlike temperate regions temperature is not a major contributing factor, except at high altitudes.

Tropical Rain Forest

Tropical rain forest is one of the oldest types of vegetation, which, theoretically, should cover much of the land lying between 10°N and 10°S. The fact that large tracts of this area are not covered with such vegetation is due to the effect of altitude, or the deflection of cloud belts by wind which causes drought at certain times of the year. Man, also, has been responsible for the extinction of much tropical rain forest vegetation.

Tropical rain forest is the climax vegetation in areas with 2000–4000+ mm of rain per year, average temperatures of 25°C with little variation throughout the year, and an average humidity of 80 per cent. The largest area of tropical rain forest today occurs in the Amazon basin of South America, while extensive tracts are also found in the Indo-Malaysian area. In Africa, this type of vegetation is found around the Gulf of Guinea and in the Congo basin.

Tropical rain forest is the most species diverse of any vegetation. Plants growing in such habitats receive continuous water and warmth, while a deficiency of nutrients is unlikely to occur due to rapid recycling. The only limiting factor is light, and this only applies to plants of the lower canopies. The herbs and shrubs found within the forest have become well adapted to growing in shade, however, and can make full use of any light that is available.

The number of tree species found in tropical rain forests is larger than that found in any other type of vegetation, and, in general, no one species is dominant. For example, parts of the Amazonian rain forest contain an average of 240 species of trees and shrubs per hectare, while new species in remote areas are continually being discovered. Some Malaysian forests, however, are dominated by members of the Dipterocarpaceae family, while members of the Leguminosae are often dominant in South American forests. Island forests can differ from those of the mainland and the tropical rain forests of the West Indies are dominated by one species – *Mora excelsa*. Single species dominant forests also occur locally inland, such as the *Cynometra alexandri* forests of Uganda and Zaïre and *Eperua falcata* in Guiana.

Because the number of tree species in most tropical rain forests is so large, individual members of any one species are well scattered.

Typical tropical rain forest consists of two or three tree layers, a shrub layer, and a herb layer. Despite the large number of tree species and the floristic differences between the forests of Asia, Africa, and South America, most trees have a monotonous, sombre appearance which is characteristic of tropical rain forests throughout the world. A typical tree has a straight, slender trunk with thin, smooth bark and branches occurring near the top of the trunk. The crowns are usually small, their shape depending on whether they are top, middle, or lower storey trees. The roots are concentrated in the surface litter and soil layer and extra anchorage is achieved with buttress roots. Leaves are dull, dark green, leathery, and more or less entire. Simple leaves predominate and even the leaflets of members of the Leguminosae have a similar appearance. Young leaves are often red or white, and nodding foliage, in which the limp, young leaves are produced on slender stems which hang downwards, is common. In general, flowers are greenish or white and inconspicuous. *Cauliflory*, in which flowers are produced on trunks and older branches, is common.

Flowering, fruiting, loss of old leaves, and growth of new leaves takes place continuously throughout the year. It is not uncommon to find two members of the same species flowering and fruiting at different times. Some trees even flower on some branches while fruiting on others.

Trees of the three strata differ not only in their height, those of the top layer reaching 40 m or more, while those of the lowest layer average only 10 m, but also in their shape. Top or A storey trees are widely scattered and, as they do not therefore compete for light with one another, they are able to develop wide, spreading crowns. The middle, or B storey trees are more closely packed, their crowns being smaller and rounded or elongated. The lower, or C storey trees often form a dense, closely packed layer, their crowns being of various shapes.

Seedlings of the upper storey trees are heliophytes and are therefore unable to develop to maturity under the normal, shaded conditions of the forest interior. It is only when a clearing appears as a result of the death of an old tree that an A storey sapling is able to reach maturity. It is characteristic of rain forest trees that saplings are generally unable to grow near a mature member of the same species. Thus the sapling replacing a dead tree is almost invariably a different species.

The shrub layer of a tropical rain forest consists of a mixture of true shrubs and saplings which are unable to mature due to lack of light, while the herb layer consists entirely of shade loving plants (sciophytes). Such plants have become well adapted to growing in the intense shade of the forest floor. Families well represented include Acanthaceae, Araceae, Morantaceae, Commelinaceae, and Zingiberaceae, with ferns and a few grasses. Plants of the herb layer are well scattered, so that it is easy to walk within the forest, although at the edges, where more light is available, impenetrable thickets of vegetation occur.

Scattered throughout the main vegetation layers of the tropical rain forest are lianas and epiphytes, which have solved their light requirement problems by climbing over other vegetation (lianas) or growing on the branches or trunks of trees (epiphytes) (Plate 6.1).

The microclimate of the tropical rain forest

The climate experienced within the tropical rain forest is not the same as that at its edges. Although average temperatures vary little throughout the year, daily fluctuations can be quite wide at tree top level, temperatures rising by day and falling by night. Within the forest, however, such variations are considerably moderated, so that the temperature experienced by the herb layer is almost constant.

Although the tree layers intercept some rain, this is not important as there is always sufficient to maintain the water balance of all plants. It is only at the extreme edge of tropical rain forest distribution that rainfall becomes important, as droughts of more than about one month cause significant changes in the species composition of the vegetation.

The relative humidity within the forest is almost constant at 80 per cent, but that at tree top level can vary from 100 per cent at night to around 60 per cent when the sun is shining. Top and second storey trees experience about five hours of bright sunlight each day, but plants at ground level within the forest receive only about 1 per cent of this light. Many rely on sunflecks for their light requirements. As light passes through the forest canopy the various wavelengths are absorbed and reflected to different extents. Thus the quality of the light reaching the forest floor is different from that entering the canopy. Plants growing beneath the canopy, therefore, have had to adapt to these spectral changes, as well as to an overall lack of illumination.

Below the forest canopy carbon dioxide concentration is higher than at tree top level and is never a limiting factor for photosynthesis. This increase in carbon dioxide concentration is due mainly to the rapid decomposition of litter on the forest floor.

Plate 6.1
Epiphytes and lianas in the tropical rain forest. (Reproduced from Plants and Environment – A textbook of Autecology by R. F. Daubenmire, 3rd edition, 1974, by permission of John Wiley & Sons Inc.)

Tropical Deciduous Forest

Tropical deciduous forest occurs on the margin of tropical rain forest and in areas with a monsoon climate, such as India, Burma, Indochina, East Africa, and North Australia. The climate of the deciduous forest region is drier than that of the tropical rain forest, rainfall averaging 1000–2000 mm per year, spread over 6–9 months. Areas with a monsoon climate experience strong winds and temperatures which depend on the season.

Tropical deciduous forest usually contains only one tree layer, with true shrubs and herbs being more numerous than in the rain forest. The annual drought, characteristic of tropical deciduous forest areas, has several effects on the vegetation. The majority of trees lose all their leaves during the dry season, the duration of leaf loss depending on the water reserves in the soil, so that this can vary in different parts of the same forest. Along river banks, where there is always ample soil moisture, leaf loss is variable. However, not every tree species is deciduous, a few hardy evergreens are also found within the canopy, where the humidity is highest. The leaves produced by these trees are quite different from those of deciduous species, being small, leathery, and often highly toxic to predators.

Many of the trees found in deciduous forests flower as the dry season ends. Transpiration through flowers is slight and thus does little to upset the water balance of the plant. However, by flowering as the rains appear deciduous trees are able to concentrate on the production of new leaves, fruits, and seeds while soil moisture is readily available.

Flowers produced by deciduous trees are often large and brightly coloured in contrast to those of the evergreens. They usually appear on the crown and outside edges of branches where they are easily seen by their animal or insect pollinators.

During the drought period tropical deciduous forest is very susceptible to fire, both natural and man-made. Therefore most of the trees have evolved thick bark which is often deeply fissured. Such trees also differ from those of the rain forest by not having buttressed roots.

The majority of herbs found in tropical deciduous forests produce their leaves and flowers during the wet season and overcome the problems of drought by lying dormant, often as bulbs, corms, or tubers (geophytes). Annuals complete their life cycle while the soil remains moist, their seeds lying dormant until the next wet season.

Montane Forest

The effects of increasing altitude on climate are responsible for the very different types of vegetation found as a mountain is ascended. In tropical regions, the heavy rainfall experienced by the lower slopes results in forest which is more luxurious than that found in the surrounding lowlands. However, on ascending the mountain the temperature falls and lowland forest is replaced by sub-montane forest, which is only two-storeyed and has fewer tree species.

Montane forests, found at cloud level, are continually drenched in moisture but experience low temperatures and light levels, due to continuous fog. The temperatures experienced by tropical cloud forests depend on altitude and govern the species of plants found within such forests. However, the temperatures within a particular forest show little fluctuation, the vegetation being exposed to an almost constant temperature and water content and air with 100 per cent relative humidity. Such climatic conditions encourage the growth of epiphytes, particularly mosses, liverworts, and ferns. The filmy, hanging tree ferns (Hymenophyllaceae) are especially characteristic of such habitats. These ferns are so dependent on water vapour that they roll up their leaves if the relative humidity falls below 100 per cent. Members of the Hymenophyllaceae belong to a group of plants known as *hygrophytes*, which can only grow in conditions of high relative humidity.

Bamboos are the dominant species of the wet, lower slopes of cloud forests, while *Selaginella* covers open ground. Very high altitudes are characterized by a single storey of stunted trees with thick, knarled trunks and many branches which are usually twisted. Such areas are often known as elfin forests.

Savannas

Trees and grasses are generally mutually exclusive, the dominance of one or the other depending mainly on available soil moisture and rainfall patterns. Grasses predominate in areas of the tropics which experience low rainfall and several months of drought, as they are better able to tolerate lack of water. In wetter areas, however, trees are dominant, as by shading they prevent heliophytic grasses from developing.

Savanna woodland occurs in areas which are too dry to support tropical deciduous forest, but which receive around 1000 mm of rain per year. The trees are widely spaced, ensuring that sufficient light reaches the ground for grasses and other herbs to grow. Xerophylous shrubs are also common, but epiphytes are rare as the atmosphere is too dry. Trees characteristic of the savanna woodland are usually small and deciduous. In general they lose their leaves during the dry season and flower at the end of this season. They are both drought and fire resistant and many belong to the Leguminosae family.

The grasses die down during the dry season, thus conserving water, but the dead leaves are easily ignited, and fires, both natural and man-made, are frequent. In fact, fire is a necessary agent for the continued existence of many savannas.

The natural scarcity of trees in savanna regions has led to the decimation of savanna woodlands by man, who uses the timber for building and fuel. Many areas which should support woodlands are now dominated by scrubby vegetation, often known as 'bush'.

The thorn woodlands of African savannas occur in areas with only 250–900 mm of rain per year and at least seven months of continuous drought. Humidity is low and temperatures fluctuate widely between day and night. Trees are scarce, those which are able to grow in such regions being well adapted to drought. *Acacia* species with their small leaflets are common, as are members of the Euphorbiaceae and Cactaceae families. The baobab (*Adansonia digitata*) is characteristic of some thorn woodlands. This tree is generally much larger than other tree species and survives drought by storing water in the swollen trunk.

The majority of the trees and shrubs of the thorn woodland have thorns or prickles which act as protective devices against the many grazing animals found in savanna regions.

In both savanna and thorn woodlands fire and the activities of man have cleared vast areas of trees, leaving a cover of grasses or sedges. Some trees, especially palms and members of the Leguminosae family, do occur but they are widely spaced and often stunted. The grasses are very vulnerable and easily succumb to overgrazing, resulting in the production of desert-like tracts of land from which the soil is rapidly eroded.

Deserts

Tropical deserts such as the Sahara or Arabian Desert are natural, but others, scattered throughout the region, are man-made, being the result of bad cultivation, overgrazing and fire. True deserts receive less than 250 mm of rain per year and it is not unusual for several years to pass with no rain falling at all. Temperatures show the widest fluctuations between day and night of any region, being very high by day, due to lack of cloud cover, and low at night, due to radiation from the bare ground. The lack of moisture results in low humidity, while hot, daytime winds quickly remove any moisture which may be precipitated as dew during the night. Thus deserts are extremely inhospitable places for plants, and only those species with extreme adaptation to drought can exist. Nevertheless, deserts are not completely devoid of vegetation (Plate 6.2), although it can be widely scattered with large tracts of bare ground. A day

Plate 6.2
Desert vegetation. (Reproduced from Plants and Environment – A textbook of Autecology by R. F. Daubenmire, 3rd edition, 1974, by permission of John Wiley & Sons Inc.)

or so after a shower of rain the ground is often covered with flowering plants, germination, flowering, and seed production taking place in the few days that the soil remains moist. The seeds of these *ephemerals* then lie dormant until the next shower, which may not be for several years.

In places where the rainfall is not so erratic geophytes grow. These also complete their life cycle during the short period when the soil remains moist and then lie dormant as tubers, bulbs, etc. until the next rain falls.

No true trees can grow under desert conditions, but tamarix shrubs and succulent members of the Euphorbiaceae and Cactaceae families occur. Tamarix roots can penetrate the soil as deeply as 50 m to find moisture in the capillary fringe above the water table. The Euphorbiaceae and Cactaceae species conserve water by having very small leaves, deciduous leaves, or no leaves at all. Their spines and stems contain abundant chlorophyll enabling photosynthesis to take place in the absence of leaves. Some plants conserve water and protect themselves from widely fluctuating temperatures by producing hairs, which can be both dense and long.

Swamp Vegetation

Swamp vegetation occurs wherever water is unable to drain away quickly, and is thus found both inland and on coastal strips. The inland swamps are usually dominated by members of the Cyperaceae, with palms, such as *Raphia* and *Raystonea* species also common. The Sudd swamps of the Sudan contain large floating islands of papyrus (*Cyperus papyrus*) and *Vassia* grass, while the South American *Eichhornia* and *Pistia* species have spread to swamps throughout the world.

Because of the rapid breakdown of litter in tropical areas, humus does not usually accumulate in swamps as it does in temperate regions. However, in some areas of South-east Asia extensive tracks of peat occur, which support a specialized peat moor forest vegetation.

Mangrove swamps occur on the shores of the tropics where outlying coral reefs moderate the force of the sea. The vegetation occurring in these swamps requires periodic flooding with sea water. Mangroves, the dominant species, are obligatory halophytes. The roots of these trees are able to filter out most of the salt and the little that does enter the plant is stored in the vacuoles and excreted through the leaves. Mangrove seedlings are viviparous, developing on the plant away from high salt concentrations. It has been found that developing mangrove seedlings contain very little salt.

The level of the soil in mangrove swamps continually changes due to sea currents and silt brought down by rivers. Mangroves have adapted to such variations in soil depth by producing stilt roots. The establishment of seedlings, however, would be extremely difficult under conditions of variable soil depth, another reason for their development on the parent plant.

The most widespread of the mangroves belong to the *Rhizophora* (Plate 6.3) and *Avicennia* genera, although species belonging to other genera also occur.

Avicennia species have the highest salt tolerance and are therefore found on the landward edges of swamps, where evaporation during periods of drought concentrates the salt. At lower salt levels *Rhizophora* species are dominant.

The establishment of mangrove swamps is an important factor in the reclamation of land from the sea.

Inland, the salt lakes of the tropics support a very specialized vegetation. Such lakes contain a high concentration of salts and the plants (halophytes) growing in such saline soils are well adapted to cope with the stresses engendered by high concentrations of ions. Many halophytes are members of the Chenopodiaceae family.

Plate 6.3
A mangrove (*Rhizophora mangle*) showing the stiltlike prop roots which enable it to grow in water. (Reproduced from Plants and Environment – A textbook of Autecology by R. F. Daubenmire, 3rd edition, 1974, by permission of John Wiley & Sons Inc.)

Suggestions for Further Reading

Ewusie, J. Y. (1980). *Elements of Tropical Ecology*. Heinemann.
Janzen, D. H. (1975). *Ecology of Plants in the Tropics*. Arnold.
Longman, K. A. and Jenik, J. (1974). *Tropical Forest and its Environment*. Longman.
Richards, P. W. (1952). *Tropical Rain Forest*. Cambridge University Press.
Walter, H. (1971). *Ecology of Tropical and Sub-tropical Vegetation*. Oliver and Boyd.
Whitmore, T. C. (1975). *Tropical Rain Forest of the Far East*. Clarendon Press.

Chapter 7

Interactions Between Plants

Introduction

Green plants are usually considered independent organisms, in contrast to other plants and animals, as they are able to synthesize their own food. In reality, however, not even green plants are truly independent, as they are reliant upon, and influenced by, many other organisms, although in ways which may not at first seem obvious. Many plants are dependent on birds, animals, or insects to pollinate their flowers and spread their seeds over as wide an area as possible. The carbon dioxide absorbed by green plants during photosynthesis was originally evolved during respiration by other organisms, while the oxygen required for respiration has accumulated in the atmosphere largely as a result of photosynthesis by previous generations of green plants. The amounts of heat, light, moisture, and nutrients available to one plant are all conditioned by the proximity of other plants. Furthermore, at least some injury from disease producing organisms and herbivores is sustained by almost all plants. The interactions between plants and animals will be discussed in the following chapter; here we consider only the ways in which plants affect one another.

Competition Between Plants

Unless conditions are very harsh, an area of bare soil will eventually be colonized by a plant community, each species filling a different niche. During the establishment of the community *competition* between plants trying to occupy the same niche occurs. Thus the physical suitability of an area for a particular species does not ensure that the species will be found growing there. This depends on the nature of the other species trying to colonize the area.

Two types of competition between plants occur – *intraspecific*, which takes place between plants of the same species, and *interspecific*, which occurs between plants of different species trying to occupy the same niche. Intraspecific competition ensures that only the fittest members of a species survive. Many factors contribute to interspecific competition, however, and the losing species may be eliminated altogether or forced into a different niche. A plant thus has a *physiological optimum*, or set of conditions under which it grows best on its own, and an *ecological optimum*, or set of conditions under which it thrives in the company of other plants. The two optima can be quite different; for example, several calcifuges are lime tolerant if grown away from other plants but are unable to compete with calcicoles in a plant community growing on alkaline soil.

The factors for which plants compete include light, soil moisture and oxygen, nutrients and carbon dioxide. External factors such as the presence of pollinators, seed dispersal agents, soil conditions, humidity, wind, and disturbance of the environment by man also affect the chances of survival of a particular species in a particular area. However, although an uninhabited area may be characterized by a set of physical parameters, as soon as it is colonized by plants and animals these parameters change. Through their modifying influence on wind, light, temperature, and humidity plants create a microclimate which may be more favourable to other species. Similarly, plants alter soil characteristics by reducing soil moisture and nutrients and adding humus. The total effect of the activities of colonizers is to create an environment in which they themselves are unable to compete with other species. Eventually these primary colonizers are eliminated from the area, resulting in a *succession* of plants. Thus a colonized area is never static but gradually changes in species composition until the climax is attained.

Because a habitat contains insufficient resources to support all species capable of living there, competition is inevitable, and the success of a particular plant depends on its ability to compete for space, light, water, and soil nutrients. In places where conditions are harsh, such as deserts and montane environments, plants are widely spaced and therefore face little competition. In the ideal conditions of the tropical rain forest, however, competition is intense and plants have had to adapt to many different niches to survive.

The rate of germination of seeds and the subsequent growth of seedlings can be the deciding factor in the ability of a particular species to overcome competition. For example, it has been found in tropical grasslands that *Andropogon* replaces lalang grass (*Imperata cylindrica*) because the former grows to a greater height and spreads more rapidly, thus obtaining a greater share of light, water, and soil nutrients.

Space is important at the seedling stage, when weaker plants can be crowded out. Competition is most intense amongst plants of the same species, so that large stands of a single species are rarely found in nature. The seed density of crop plants must be carefully calculated to ensure maximum number of plants per unit area with minimum competition. Only thus can maximum yield be obtained.

In the tropical rain forest it is found that mature trees depress the development of seedlings of their own species, but seedlings of other species are able to grow in close proximity, providing other factors do not interfere. This is an important factor in the maintenance of the species diversity so characteristic of the tropical rain forest ecosystem.

Above ground, light is the most important factor in competition between plants. In fact it has been suggested that all above ground competition ultimately depends on light. As light cannot be stored it must be used at maximum efficiency and this is achieved in nature by the large range of light requirements shown by different species. Sun and shade plants can live in close proximity as they fill different niches, but competition occurs between heliophytes and between sciophytes trying to occupy the same area. Competition for light is the reason for the complicated structure of the tropical rain forest, an ecosystem in which the maximum use of light occurs (fig. 7.1).

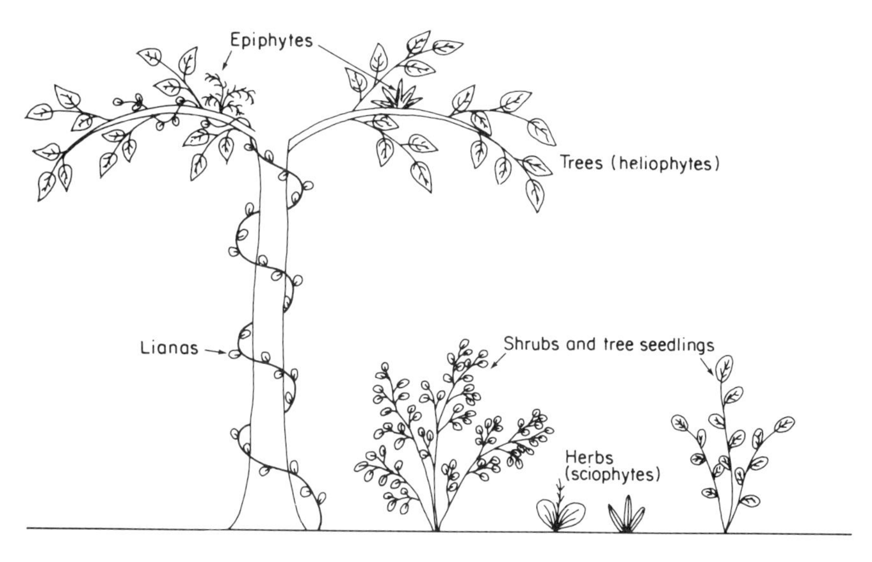

Fig. 7.1
Structure of the tropical rain forest

Competition for light occurs between the leaves of a single plant as well as between different plants. A leaf is a self-contained entity and if unable to photosynthesize it dies. Thus the lower leaves of a broad leaved plant are at a distinct disadvantage.

Competition between plants for carbon dioxide must occur, especially in crowded areas with a high light intensity, but such competition has been little studied. The percentage of carbon dioxide in the atmosphere today is lower than it has been in the past and is well below the optimum for maximum photosynthesis. The artificial application of the gas to plants grown under glass usually increases growth and yield. Tropical plants with a C_4 pathway (see Chapter 4) are at a distinct advantage over C_3 plants as they do not lose carbon dioxide through photorespiration. CAM plants, in absorbing carbon dioxide at night when other plants close their stomata, must be the most successful in competing for the gas.

The percentage of oxygen in the atmosphere today is high compared with that present when plants were first developing chlorophyll-based photosynthetic mechanisms. Thus there is little or no competition for this gas above ground, although in the soil it can be the limiting factor for the growth of some plants (see Chapter 5).

Below ground, plants compete for water, air, and nutrients, their ability to obtain these essential requirements depending on the rate of root growth. This in turn depends on photosynthetic ability, so that it is not possible to separate above and below ground factors in competition, as each influences the other (fig. 7.2). An inability to compete for nutrients results in reduced shoot growth, the plant eventually being shaded out by more vigorous neighbours. But lack of light reduces root growth and therefore the ability to compete for soil water, air, and nutrients. Thus a vicious circle is established which leads eventually to the death of the plant. For a plant to survive in a crowded habitat, such as the tropical rain forest, it is imperative that it germinates and grows quickly, for it is at the seedling stage that competition is most intense.

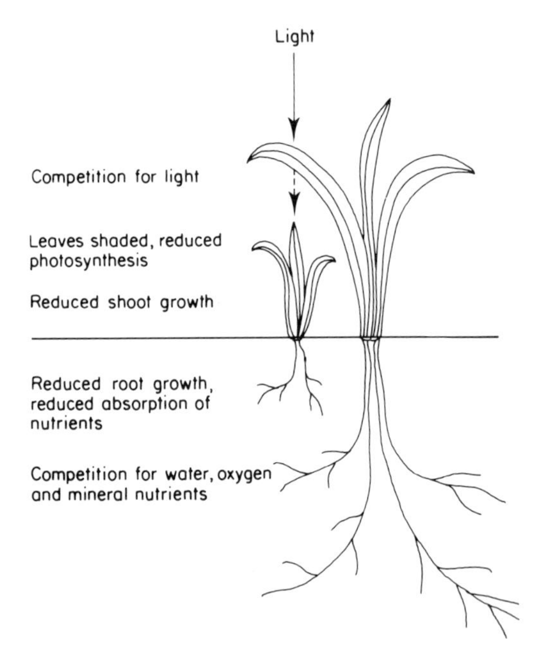

Fig. 7.2
The effects of competition on the growth of plants

The ability of plants to compete depends to some extent on the nutritional status of the soil. In a soil with a high concentration of calcium ions, calcicoles grow at the expense of calcifuges, but the reverse is true in soils with a low calcium concentration. Many plants growing in unfavourable habitats survive because they have adapted to adverse conditions, such as lack of water or soil nitrogen or the presence of toxins in the soil. Drought tolerant plants are discussed in Chapter 3 and those specifically adapted to withstand high levels of toxic metals, such as copper, nickle, and aluminium are described in Chapter 2.

In more favourable environments such plants are unable to compete and their niche is filled by other plants.

Plants able to grow in close proximity avoid competition for soil water and nutrients by rooting at different levels, examples being the grasses and shrubs found in the tropical savannas. The grasses are comparatively shallow rooted plants, while the shrubs often have long tap roots reaching to the capillary fringe (fig. 7.3).

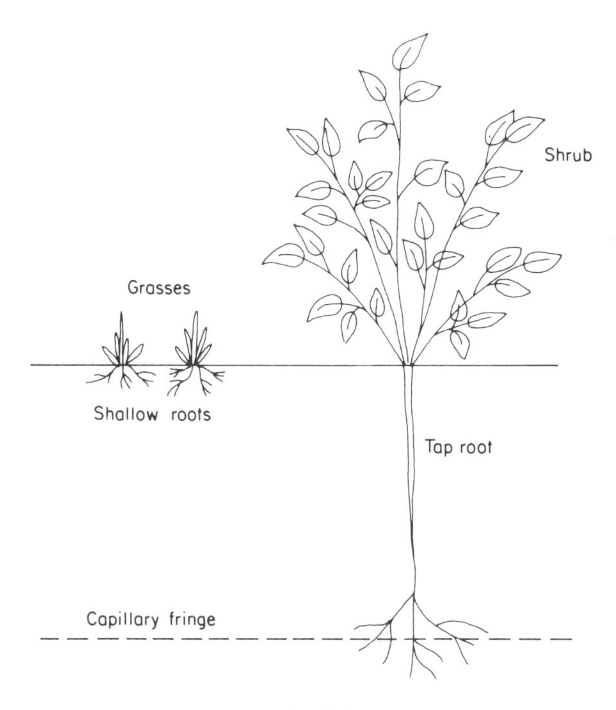

Fig. 7.3
An example of plants which do not compete for water

Symbiosis

The term *symbiosis* means 'life together' and this includes all the effects of one organism on another. In this chapter only the effects of one plant on another will be considered, the effects of animals on plants are described in Chapter 8.

Symbiotic effects can be loosely considered under three headings – *commensalism*, in which one plant benefits without actually harming the other; *mutualism*, in which both plants benefit, and *parasitism*, in which one plant benefits at the expense of the other.

Commensalism

The many lianas and epiphytes found in tropical forests are examples of commensalism. They use other plants for support, but in general do not harm them, except possibly by shading. Lianas are rooted in the ground but their stems need the support of other plants in order that the leaves receive maximum light. Although small lianas seldom harm their hosts, larger species can cause death through smothering.

Epiphytes also use other plants for support in order to obtain light, but they are not rooted in the ground. Because epiphytes are usually so much smaller than their hosts they seldom do any harm, although they have been known to cause the deaths of small cocoa and citrus trees through smothering.

As the abundance of lianas varies directly with warmth and humidity, they are a characteristic feature of the moist tropics and can cause considerable interference in forestry. Lianas can be classified as follows:

1. Leaners – plants which have no special device for holding onto a support, e.g. *Plumbago capensis*.
2. Thorn lianas – plants producing thorns or prickles, which although not specifically produced for the purpose do help to anchor the liana to a support, e.g. *Bougainvillea* spp.
3. Twiners – plants, mostly herbaceous, in which the whole stem twines around the support, e.g. *Ipomoea* spp.
4. Tendril lianas – plants possessing special organs, the tendrils, which are specifically produced to help the liana climb over its support, e.g. members of the Cucurbitaceae and Leguminosae families.

Lianas may also be classified as heliophytes or sciophytes. The former spread their foliage over the canopy of the supporting tree or shrub, but sciophytic lianas, such as *Monstera* and *Vanilla* only climb the stems of their hosts and complete their life cycle without ever reaching the sunny surface of the canopy.

Epiphytes include cryptograms, herbs, shrubs, and trees; about 33 families of flowering plants have epiphytic species, the majority belonging to the Araceae, Asclepiadaceae, Bromeliaceae, Cactaceae, Orchidaceae, and Rubiaceae families. Many ferns are epiphytic. Epiphytes are found growing on trees, shrubs, lianas, and submerged plants. Often a specific epiphyte shows a marked preference for a particular supporting species, while some trees support more epiphytes than others, depending on the texture and chemical composition of the bark. On woody plants (Plate 7.1) epiphytes may perch on the trunk, branches or on the upper surfaces of evergreen leaves. When in the last position they are known as *epiphylls*. Epiphytes are especially abundant in the forks of trees and on horizontal branches, where anchorage is easiest and soil can collect. Epiphytes are least abundant on vertical and smooth surfaces. The oil palm (*Elaeis guineensis*) is an excellent host, as the persistant leaf bases provide pockets for anchorage and the collection of soil and moisture. It is unusual to find a wild oil palm which is not festooned with epiphytes, although when grown on plantations these invaders are kept to a minimum to increase yields.

By growing on other plants epiphytes obtain vital light, but to do this they

Plate 7.1
The development of a strangling fig on a palm tree. Eventually the fig will kill the palm. (Reproduced from Plants and Environment – A textbook of Autecology by R. F. Daubenmire, 3rd edition, 1974, by permission of John Wiley & Sons Inc.)

sacrifice water and mineral nutrients. Of all classes of vegetation epiphytes are the most dependent on precipitation, so that these plants are most abundant where drought is never protracted. In cold, dry climates epiphytes are restricted to algae, lichens, liverworts, and mosses, but in warm, wet climates ferns and flowering plants abound.

In the tropical rain forest, where epiphytes are most abundant, these plants cover a wide range of the ecological spectrum, varying from hardy, drought resistant heliophytes to sciophytes, which, growing within the canopy, seldom lack moisture and a humid atmosphere. Epiphytes growing on the tops of trees include cacti and bromeliads which are specifically adapted to withstand intense radiation and drought. The most abundant group are the heliophytes which grow within the crowns of trees and on large branches. They sacrifice some light for a more equitable microclimate. The sciophytes occur well within the canopy on the trunks and lower branches of trees and shrubs.

Epiphytes obtain nutrients from rain water and debris which collects in cracks and hollows on the surfaces of tree trunks and branches. Ants and other organisms help to break down the debris to release nutrients.

In order to survive when little water is available many epiphytes show xeromorphy, including thickened cuticle, sunken stomata, and succulence. Some epiphytes spread their roots over the plant surface in such a way that they are able to absorb the maximum available water. The nest epiphytes develop a dense mass of roots, which look like a bird's nest. These collect debris and hold water so that eventually the nest holds a pocket of soil. Tank epiphytes (Plate 7.2) have long, broad, stiff leaves which form a rosette for storing water. Insects, including mosquito larvae, live in the water, while others fall in and drown, providing nutrients for the plant. Such plant pools are a microcosm, as they often contain organisms found nowhere else. Tank plants belong to the Bromeliaceae family and are common in South American forests.

The roots of many tropical epiphytes belonging to the Orchidaceae and Araceae families extend outward into the air, appearing as thick, unbranched,

Plate 7.2

A tank epiphyte. (Reproduced from Plants and Environment – A textbook of Autecology by R. F. Daubenmire, 3rd edition, 1974, by permission of John Wiley & Sons Inc.)

whitish organs. Covering the surface of these roots is a special layer of cells which can take up water rapidly from the briefest of showers. Afterwards the living core of the root absorbs the water from this storage layer, which is known as *velamen*.

Among the epiphytic species of *Tillandsia* (Bromeliaceae) (Plate 7.2) the roots serve chiefly as anchorage organs, the leaves and stems taking over the function of absorption. In Spanish moss (*T. usneoides*) the finely divided shoot system is covered with scales that collect water. This is absorbed by the small, uncutinised spots on the epidermis which are sheltered by the scales during times of drought.

Some aerial plants, known as *hemi-epiphytes*, produce long roots which eventually reach the ground and become anchored in the soil. At this stage the plant ceases to be an epiphyte as it is no longer entirely dependent on its host for support.

The colonization of an area with epiphytes begins with algae, and succession through lichens, mosses, and liverworts occurs until an environment suitable for ferns and flowering plants is established.

Mutualism

Mutualism occurs when two plants growing in close proximity benefit each other. Lichens have developed mutualism to such a high degree that the partners are unable to exist on their own. Each lichen species contains a fungal component and an algal component, the cells of the alga being enmeshed within the mycelium of the fungus. Reproduction takes places through *soredia*, which are pieces of fungal tissue enclosing a few algal cells. The green or blue–green algal component produces carbohydrates for the whole plant through photosynthesis, while the fungal component decomposes litter to release mineral nutrients for the lichen. The blue–green algal components of some lichens are also able to fix nitrogen.

The partnership of fungi and algae in lichens is so successful that these plants are able to colonize areas where the conditions are too harsh for other plants to grow. Lichens are long lived and able to endure extremes of temperature and water supply and low availability of nutrients. The products of photosynthesis are stored as sugar alcohols in the fungal component, so that lichens can remain alive during long periods of desiccation and on rewetting rapidly resume photosynthesis and nutrient uptake.

The bacteria living in the leaf nodules on *Ardesia* and *Psychotria* species are another example of mutualism, as they benefit the plant, probably through the production of a growth hormone.

Mycotrophy

Mycotrophy, the association of fungi with the roots of higher plants, is an example of mutualism, as the fungus acts as an intermediary in the uptake of nutrients by the root. The fungal mycelia combine with the roots to form a compound structure known as a mycorrhiza. Mycorrhizae are of two chief types, although integration between these extremes occurs widely. *Ectotrophic* mycorrhizae result when the mycelium forms a dense mantle over the surface of the root, with many hyphae extending outward into the soil and others penetrating the root by forcing their way between cells. However, the majority of mycorrhizae are *endotrophic*, some hyphae living in the protoplasts of root tissue while others extend outward into the soil, but without forming a surface mantle. Members of the Pinaceae family have ectotrophic mycorrhizae, while orchids (Plate 7.3) and members of the Compositae are endotrophic.

Orchid seeds are extremely minute, containing only the rudiments of an embryo with a small amount of fatty reserves, and they will not germinate normally unless in the presence of the mycelium of a fungus such as *Rhizoctonia*. However, it has been found that plants can be grown non-symbiotically from seed if sugar is supplied and the pH regulated to 5 or lower.

Many rain forest trees have virtually no root hairs and are dependent on fungi which are specialists in gathering minerals. The fungus obtains these nutrients from decaying litter and humus. In exchange, the tree provides the fungus with carbohydrates, which are absorbed through the hyphae which penetrate the roots. Fungi have no chlorophyll and are therefore unable to photosynthesize their own carbohydrates.

Plate 7.3

The reduced root system of an orchid which relies on mycotrophy for nutrient absorption. (Reproduced from Plants and Environment – A textbook of Autecology by R. F. Daubenmire, 3rd edition, 1974, by permission of John Wiley & Sons Inc.)

Nitrogen fixation

The nitrogen-fixing activities of bacteria and algae which live in the soil, on plant surfaces or within a plant host are of extreme importance in maintaining the nitrogen cycle described in Chapter 1. The substrate for these organisms is the nitrogen gas of the air, a very unreactive substance, which nevertheless some microorganisms are able to reduce to ammonia. Nitrogen-fixing microorganisms usually grow in association with plants, so that the ammonia is rapidly absorbed by the plants and converted first to amino acids and then to a variety of nitrogen-containing organic compounds, the most important being proteins.

The enzyme nitrogenase which catalyses the reduction of nitrogen to ammonia is sensitive to oxygen. It is probable, therefore, that the main advantage to a microorganism in living within a plant is that anaerobic conditions can be maintained. As nitrogen fixation only takes place under low soil nitrogen conditions, the advantages to the plant in such an association are obvious. When growing in environments containing low concentrations of soil nitrogen plants acquiring their nitrogen from associated microorganisms are at a distinct advantage. They are likely to grow faster and are thus better able to compete than plants without such associations.

Although several free-living, nitrogen-fixing microorganisms exist their contribution to soil nitrogen is small, and it is only when nitrogen-fixing organisms are associated with plants that their contribution to the overall nitrogen budget becomes important.

The roots of many leguminous plants have nodules inhabited by bacteria (*Rhizobium* spp.), which fix atmospheric nitrogen in soils deficient in nitrogen. The nitrogen eventually benefits the plant, while the bacteria obtain carbohydrates and water from their hosts. The co-existence of plant and bacteria is thus an example of mutualism. About 80–90 per cent of the species of the sub-family Papilionaceae are associated with nitrogen-fixing bacteria, while only 25 per cent of Mimosaceae and very few Caesalpiniaceae form nodules.

When the host plant seedling starts to produce leaves it excretes a substance which attracts the nitrogen-fixing bacteria. The bacteria, in turn, excrete a hormone which enables them to penetrate the root hairs and spread into the root, where they multiply, forming nodules. There are many *Rhizobium* species, each associated with a particular group of higher plant species. It is interesting to note that the plant will only secrete the attracting substance if the soil is nitrogen deficient.

Nodules which are actively fixing nitrogen are usually red, due to leghaemoglobin, a protein produced by the plant. Leghaemoglobin supplies oxygen to the bacteria for respiration without affecting nitrogenase.

Although nitrogen fixation associated with legumes has been the most widely studied, at least 120 species of non-leguminous plants, mostly trees and shrubs, are now known to be associated with nitrogen-fixing bacteria. For example the tropical *Trema aspera* can act as host to the *Rhizobium* bacterium. Such associations are important to the nitrogen economy of tropical forests whose soils are low in nitrogen compounds.

The association of nitrogen-fixing blue–green algae with plants is also important. For example, the association of the water fern *Azolla* with the blue–green alga *Anabena* is critical to the cultivation of paddy rice. Other tropical plants associated with blue–green, nitrogen-fixing algae include the cycad *Macrozamia*, and *Gunnera*, both of which act as hosts to the microorganisms.

The association between grasses and nitrogen-fixing microorganisms is much looser. An *Azobacter*, for example, associated with the tropical grass *Paspalum notatum*, forms a sheath over the roots in which it lives and fixes nitrogen. Nitrogen-fixing bacteria also live alongside fungi which are instrumental in the decay of timber. The bacteria supply nitrogen to the fungi, while themselves obtaining carbon compounds produced during the breakdown of cellulose.

Parasites

Parasitic plants obtain all or part of their nutrients and water from other plants. To enable them to do this they have specialized roots and other organs known as *haustoria*. Pathogenic fungi and bacteria are total parasites which damage their hosts by consuming tissues and releasing toxins. However, generally a careful balance is maintained between parasite and host, ensuring that although the host is weakened, its life is not threatened. A dead host is of no use to a parasite. Should this balance be upset through external interference, however, the results can be disastrous for the host, and ultimately for the parasite. The inadvertent introduction by man of parasitic fungi, bacteria, and higher plants to new hosts which have no resistance to the parasite has led to the complete failure of crop plants, and thus to widescale starvation of those dependent on the crop for food.

Many higher plants are only semi-parasitic, as they obtain water and mineral nutrients from their hosts, but contain chlorophyll and are thus able to photosynthesize their own carbohydrates. Examples include the many members of the Loranthaceae family known as mistletoes, species of which occur throughout the world. In the tropics mistletoes are widespread, semi-parasites on trees, including cocoa. In general they do little harm to the tree unless particularly vigorous or present in large numbers. Although small trees can be killed, it has been found that cocoa trees with some mistletoe parasites are better able to withstand drought than those with none. Thus it would appear that in this case some degree of mutualism exists between parasite and host.

Striga is a genus of Asiatic annual herb parasitic on the roots of grasses. Although they possess a few green leaves, and are thus able to photosynthesize, in general their presence reduces the yield of tropical economic grasses, such as sorghum. However, since *Striga* seeds are stimulated to germinate by contact with the roots of species they cannot parasitize, crop rotation helps to control the parasite.

Members of the Orobanchaceae family known as broomrapes are widespread herbs which are complete parasites on the roots of higher plants. Broomrape roots are connected to the roots of their host plants and in some cases the seeds will not germinate unless in contact with the root of a suitable host. The aerial parts lack chlorophyll and consist of little more than a brownish inflorescence. This family is closely related to Scrophulariaceae, which contains *Striga* and many other root parasites.

Rafflesia is a genus of Malaysian plants parasitic on the roots of *Vitis*. These parasites have become so extremely degenerate that they resemble fungi, for the vegetative parts are similar to mycelia and are wholly contained within the roots of the host. One species, *R. arnoldii*, is famous for bearing one of the largest flowers in the plant kingdom. It is about 1 m in diameter and has an evil odour.

Allelopaths, Antibiotics and Phytoalexins

Plants are able to protect themselves from competitors and invaders by the production of chemicals which are toxic in some way to other plants. Some release chemicals into the environment which prevent the growth of other plants within their immediate vicinity. Such substances are known as allelopaths, and although only well documented for a few species they are probably widespread. Allelopaths can be volatile substances, such as the monoterpenoids cineole from *Eucalyptus* species and camphor from the camphor tree (*Cinnamomum camphora*). Such substances are released into the air, especially under the high temperatures experienced in the tropics, and enter other plants by dissolving in the cutin of the leaves. The growth of neighbouring plants can be affected for distances of as much as 10 m.

Many plants contain harmless glycosides, which when washed out of the leaves and into the soil are hydrolysed to allelopaths. Although these substances are soon destroyed by soil organisms they can have a harmful effect on other plants, retarding growth and preventing germination. Examples of such allelopaths include quinones, which are cell toxins, and phenolic acids. Juglone from walnut trees (*Juglans* spp.) is a well documented allelopath which prevents the growth of many species under the canopy of the tree.

Plants also produce chemicals which protect them from invasion by fungi, bacteria, and viruses. All such substances can be classed as *antibiotics*, a term which means 'against life'. Probably the best known antibiotic is penicillin produced by the *Penicillium* species of mould.

Invasion of higher plants by fungi is ubiquitous and thus during their evolution such plants have developed many biosynthetic pathways for the production of antifungal agents. 3,4-Dihydroxybenzaldehyde, for example, protects the Cavendish banana against the fungus *Gloeosporium musarum* which causes ripe fruit rot. The lignan hydroxymatairesinol, present in the bark of some trees, protects against fungal invasion, while the tannins and coumarins found in many plants throughout the world also have antifungal properties.

Other compounds are only produced by plants after fungal attack has taken place. These substances are known as *phytoalexins* and they are usually highly effective in preventing the spread of fungal invasion. Examples include orchinol from orchid species, ipomeamarone from sweet potato (*Ipomoea batatas*) and 6-methoxymellein from carrots (*Daucus carota*).

Suggestions for Further Reading

Etherington, J. R. (1975). *Environment and Plant Ecology*. Wiley.
Grime, J. P. (1979). *Plant Strategies and Vegetation Processes*. Wiley.
Harbourne, J. B. (1977). *Introduction to Ecological Biochemistry*. Academic Press.
Mishustin, E. N. and Shil'nikova, U. K. (1971). *Biological Fixation of Atmospheric Nitrogen*. Macmillan.
Postgate, J. (1978). *Nitrogen Fixation*. Arnold.
Scott, C. D. (1969). *Plant Symbiosis*. Arnold.
Vickery, M. L. and Vickery, B. (1981). *Secondary Plant Metabolism*. Macmillan.

Chapter 8

Plants and Animals

Introduction

Plants and animals depend on one another to such an extent that a macrosymbiotic relationship can be said to exist between these two types of living organisms. Such a relationship ranges from complete mutualism to complete parasitism. Primarily, plants are important to animals as a source of food. However, they also serve as protection against predators and adverse environmental conditions, and provide materials for nest and other home building, important factors which are often overlooked. By modifying the environment plants are to a large extent responsible for the formation of the various types of habitat occupied by specific animals.

Animals are less obviously important to plants, but many play an essential role in pollination and seed dispersal, especially in the tropics. Animals are also part of the *biogeochemical cycles* (fig. 8.1) described in Chapter 1, essential to life on earth, as they ensure that soil nutrients, carbon dioxide, and nitrogen are always available to plants.

Grazing and Browsing

Plants are important to animals as food, as all animals obtain vital sugars, proteins, fats, and vitamins either directly or indirectly from plants. Plants are the only means by which animals can obtain the energy necessary for their life processes to continue. Although carnivorous animals obtain their requirements from other animals, which themselves may be carnivores, all food chains start first with plants and then with herbivores as shown in fig. 8.2.

Grazing herbivores eat the leaves of grasses and other herbs, while browsing animals eat the leaves of woody plants. Antelope, zebras, etc. are grazers, while giraffes are browsers. However, the term grazing is frequently used to describe both types of food collection. Some insects graze; locusts and caterpillars, for example, eat the leaves of plants. However, many insects have a more sophisticated method of feeding. Aphids and other bugs pierce the surface of leaves and feed directly on the liquid contents, which have a much higher nutritional value than the whole leaf. Some insects even live beneath the leaf surface, gaining protection as well as food. Such a life style has been perfected

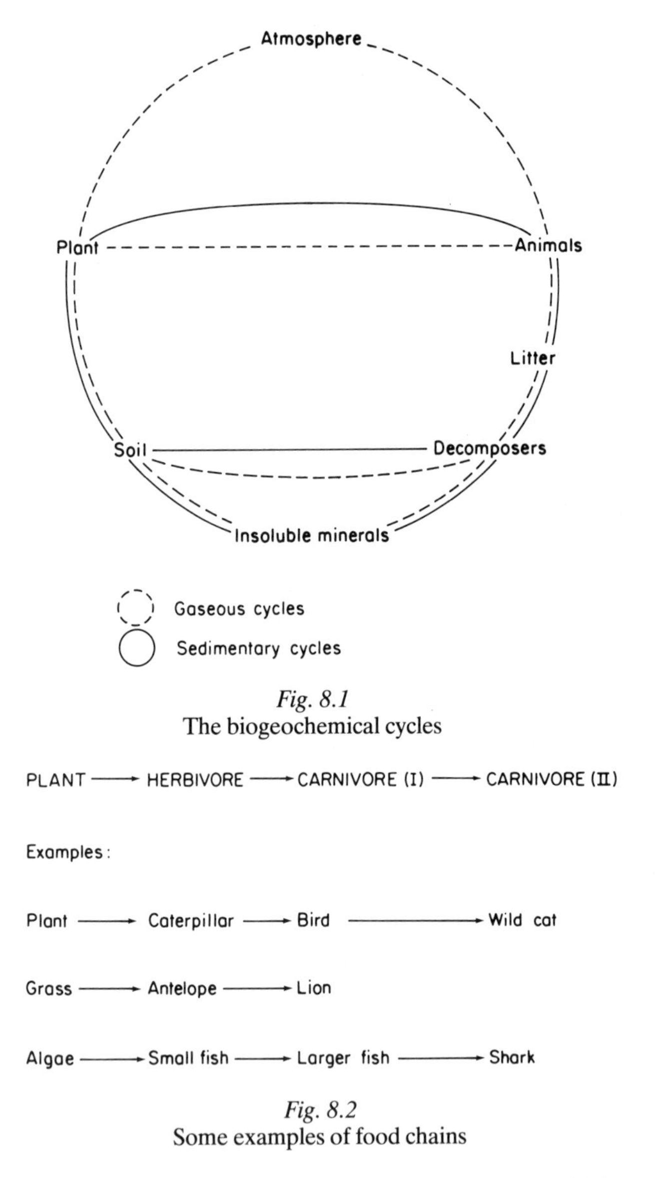

Fig. 8.1
The biogeochemical cycles

Fig. 8.2
Some examples of food chains

by those insects which live in the outgrowths, known as *galls*, which some plants produce as a response to insect invasion. Inside the galls the insects are able to obtain food of a higher nutritional value than that of normal plant sap.

Ant Plants

The relationship between some species of ant and their host plants is particularly mutualistic. In the tropics there are ants which live in the swollen bases of thorns on *Acacia* trees. These ants have very vicious stings and protect the tree against most invaders, both animal and other plants. In return the tree provides a home for the ants and food produced by the nectaries. These *Acacia* species do not need to waste their resources on producing the protective cyanogenic glycosides biosynthesized by species which are not inhabited by ants. However, it has been shown that the dependence of the tree on ant protection is absolute. If the ants are removed the tree soon succumbs to animal predation or is smothered with lianas and epiphytes.

In South America, the leaf-cutting ants gather pieces of leaf from plants and take them back to the nest, where they are used to cultivate a fungus. The fruits of the fungus produce a specially nutritious food for the ant larvae. In order to improve yields the ants add a growth hormone (auxin) to their fungal gardens. Both ant and fungus are totally dependent on one another. The fungus has never been discovered growing outside such ant gardens.

Plant Defence Mechanisms

If herbivores were able to graze plants without restraint many species, unable to reproduce sexually or vegetively, would quickly become extinct. Thus over the millions of years of co-evolution of plants and animals, including insects, there has emerged a regulatory mechanism which ensures that neither the plant nor the animal experiences such a fate. Some plants produce thorns, spines, prickles, or stinging hairs as a defence against invaders. Dense hairs on the surface of a plant also play a protective role, as insects are unable to penetrate to the leaf surface. Some hairs secrete sticky substances (gums) which immobilize the insect so that it dies. The barks of some trees also exude gum when damaged, trapping and killing the invader. Dead insects are often found inside the amber droplets collected from *Acacia* trees. The production of resins and latex by some plants is probably a similar defence mechanism. However, for almost all plants there will be at least one animal which has been able to overcome the defence mechanism (Plate 8.1).

Secondary plant compounds

The most widespread defence mechanism amongst the angiosperms is the production of secondary compounds which are toxic or act as feeding deterrents to herbivores. Primary compounds, such as sugars, proteins, and fats are the building blocks of plants and the food of animals and are thus usually harmless, but the major role of secondary compounds is to protect the plant from attack by herbivores and plant parasites. Such compounds are very diverse chemically and include alkaloids, non-protein amino acids, cyanogenic glycosides, cardiac glycosides, toxic saponins, and terpenoids, and a variety of other compounds.

Alkaloids are common in tropical plants, examples being strychnine in *Strychnos* species, nicotine in *Nicotiana* species, caffeine in coffee (*Caffea* species), and quinine in *Cinchona* species. In small quantities many of these

Plate 8.1
Opuntia infestation was controlled by introducing the cactus moth to Australia. (Reproduced by permission of Department of Lands, Queensland, Australia.)

compounds are used medicinally as they are toxic to the microorganisms causing many diseases. Alkaloids act as toxins by binding to the protein of enzymes, thus preventing them from catalysing the biochemical reactions necessary for life to continue.

Cardiac glycosides, such as ouabain from *Acocanthera* and *Strophanthus* species, act on the heart. Ouabain is used medicinally to cure some forms of heart disease, but in quantity it is fatal.

Cyanogenic glycosides release toxic hydrogen cyanide when plants containing these compounds are eaten by animals. Non-protein amino acids often act as toxins because they replace essential amino acids in proteins.

Tannins are common secondary plant compounds. Although not as toxic as the compounds described above, tannins make vegetation highly indigestible by binding to proteins.

Although the plant world as a whole produces such a spectrum of toxic compounds, an individual species usually only produces one type. Even whole families of plants can be characterized by the type of secondary compound their members produce. Probably every member of the Cruciferae, for example, synthesizes glucosinolates, the compounds which produce toxic mustard oils when the plants are eaten. Alkaloids of particular types are characteristic of particular families, for example the opium alkaloids in the Papaveraceae and the complex indole alkaloids in the Rubiaceae.

To some extent the production of secondary compounds can be correlated with habitat. Rain forest trees, for example, are richer in such compounds than deciduous trees, while plants of arid habitats are often highly poisonous, members of the Euphorbiaceae, many of which have toxic latex, being good examples.

It would seem that the production of secondary compounds is a drain on plant resources and that any one species cannot afford the biosynthetic pathway of more than one type of compound. Such specialization has been to the detriment of plants, however, as insects, their main predators, have evolved mechanisms to overcome the toxicity of certain compounds. Detoxifying mechanisms are just as energy consuming for insects, however, as the production of toxic compounds by plants. Thus any one species of insect can only detoxify one type of compound, and, therefore, feed on one or at most a few species of plants. Brucid beetles, for example, are locally host specific, the larvae of each species feeding on the seeds of a different host plant. Because host species remain toxic to other insects they are not overgrazed and the balance between plants and insects is maintained.

Because of this inbuilt ability to evolve detoxifying mechanisms it is not surprising that so many insects have become resistant to man-made toxins.

Larger animals usually eat small quantities of many plants so that they do not imbibe enough of any one toxin to do harm. In times of famine when choice of plants is limited, however, such animals can be poisoned by eating too much of a particular toxic plant. *Datura* species, for example, contain toxic atropine-type alkaloids. Such plants are only eaten by herbivores when they have exhausted more palatable vegetation.

Ruminants, a special type of herbivore, are well adapted to eating toxic plants as they host a wide variety of detoxifying microorganisms.

Toxic compounds which are also feeding deterrents often have a bitter taste, examples being alkaloids, which are found in many plant families, and cucurbitacins, which are characteristic of the Cucurbitaceae. Other feeding deterrents, such as the ubiquitous tannins, have a bitter taste, but they are not toxic. One of the most successful feeding deterrents is the compound azadirachtin, produced by the tropical neem tree (*Azadirachta indica*). This compound even deters the desert locust which eats almost every other plant within its vicinity.

Some plants produce compounds which, when eaten, interfere with reproduction or development of animals, thus effectively reducing the population of the predator. Probably because they are the oldest plants and their co-evolution with insects has occurred over such a long time, many ferns and gymnosperms produce compounds of this type. Insect predators of ferns are rare, unlike the seed-bearing plants which have not yet found a completely effective form of protection. The insect molting hormone analogues, known as ecdysones, are particularly characteristic of ferns. Such hormones control the molting of the outer skin of the insect larva and are produced in carefully controlled quantities at each stage of growth. The absorption of extra amounts of such hormones from plants, however, causes growth abnormalities and sterility.

Many leguminous plants, which are heavily grazed by animals other than insects, produce isoflavonoids, which mimic sex hormones and cause infertility or abortion of the young. Such compounds occur in some forage plants such as clovers and alfalfa. Sheep are particularly susceptible to their effects.

Insect juvenile hormone analogues are also produced by some plants, these effectively preventing the larva reaching the adult stage and reproducing. Other compounds have the opposite effect, causing metamorphosis to occur too soon, which also leads to sterility. As with other toxic compounds, however, some insects have developed detoxifying mechanisms or have evolved ways of using the compounds to their own advantage.

Some insects use plant toxins as defence materials for themselves. The monarch butterfly retains the cardiac glycosides it ingests when the larva eat milkweed (*Asclepias* spp.) so that the adult butterfly is highly toxic and therefore avoided by birds. These cardiac glycosides also have an unpleasant taste and therefore act as feeding deterrents. The aphid which feeds on oleanders (*Nerium oleander*) becomes highly toxic from the cardiac glycosides it ingests. Unlike other aphids, which are green to conceal their presence from predators, the oleander aphid is bright yellow, the danger colour of the animal world.

Many insects have turned the production of toxic secondary compounds by plants to their advantage in other ways. Volatile compounds, such as terpenoids and mustard oils, are feeding attractants for those insects which feed on the plant, although deterrents for other insects. For example, the larvae of Pierinae butterflies feed on members of the Cruciferae and Capparidaceae families which contain glucosinolates. These compounds release volatile, toxic mustard oils when the plant is damaged and the female butterfly will only lay her eggs on plants with an odour that she recognizes. Different species of butterfly recognize different mustard oils and thus no one species of plant is overgrazed, while the larvae also face little competition. Other plants produce compounds which are feeding deterrents to butterflies, examples being members of the Rubiaceae, which biosynthesize alkaloids, and members of the Ranunculaceae which contain protoanemonin.

Grasses

Only a few grasses produce toxins, usually cyanogenic glycosides, yet these plants are the food of most large herbivores and would thus be expected to have developed efficient defence mechanisms. During their co-evolution with animals, however, grasses have developed a method of growth in which grazing by animals is a positive advantage. Unlike most other plants the growing points of grasses are just below ground level and are therefore not damaged by trampling or when the shoots are eaten. In fact, grazing stimulates the rapid production of new shoots and also favours *tillering*, which is the production of lateral shoots from axial buds at the base of older shoots. Thus moderate grazing encourages the growth and spread of grasses. Should the plant be damaged the rapid growth of numerous fibrous, adventitious roots ensures that recovery is rapid.

However, even grasses can be overgrazed and eventually killed, a situation which can turn tropical grasslands into deserts, and one which leads to considerable erosion. Introduced animals usually cause more damage than indigenous species. The large herds of wildebeest of East Africa, for example, naturally migrate from one pasture to another every year, allowing regeneration to occur. In the past, tribes such as the Masai were nomadic, moving their cattle and goats regularly so that grazed pastures could rest. Increases in populations amongst such tribes and the present tendency to settle in one place have led to serious overgrazing in some areas. Large tracts of land have been, or are in the process of being, turned into deserts by such practices.

The Destructive Activities of Animals

The delicate balance between plants and animals is easily upset and many animal activities are harmful to plants. Elephants, baboons, and other large animals cause physical destruction of plants. Elephants, owing to their large size and strength, are particularly destructive, as they tear up and knock down trees. Large herds of these animals can eventually turn a forest into grassland or even desert.

Swarms of locusts can completely defoliate large areas of vegetation, causing the death of many plants, although others eventually regenerate. It is interesting that the neem tree is never eaten by locusts, because the secondary compound, azadirachtin, produced by this tree acts as an efficient feeding deterrent.

Quelea birds, which occur in large flocks in Africa, can devour whole crops of grain, while rodents devastate peas, beans, cereals and groundnuts. Smaller animals and insects, such as mites, weevils, aphids, and nematodes, etc. also damage plants. Their effects can be particularly devastating to crop plants.

Animals are also often vectors in the transmission of plant diseases. The mealy bug on cocoa trees, for example, transmits the virus causing swollen shoot disease from one tree to another.

Carnivorous Plants

Although in the majority of plant–insect relationships it is the insect which is the predator, a few plants have evolved mechanisms whereby they are able to prey on insects. Pitcher plants (Plate 8.2) are species belonging to the genera *Darlingtonia*, *Nepenthes*, *Sarracenia*, etc., the leaves of which hold a small pool

Plate 8.2

Pitcher plants which feed on insects. (Reproduced from Plants and Environment – A textbook of Autecology by R. F. Daubenmire, 3rd edition, 1974, by permission of John Wiley & Sons Inc.)

of liquid. This liquid is partly or entirely excreted by the leaf surfaces and is an aqueous solution of proteolytic enzymes. Insects and other small animals falling or wandering into the pool are unable to escape and eventually drown. The enzymes dissolve the soft parts of the bodies releasing amino acids which are absorbed by the plant. Other insects have developed ways of resisting the enzymes and are thus able to live in the pitcher plant pools, their excretory products providing nourishment for the plant.

Sundews (*Drosera* spp.) are widely distributed in boggy habitats. The leaf blades are covered with gland-tipped hairs, each of which bears a droplet of sticky liquid. Small insects alighting on the leaf become stuck and the stimulus causes more hairs to bend over and touch the body. The glands then excrete proteolytic enzymes which digest the insect. It is interesting to note that material which does not contain protein fails to stimulate hair movement or the excretion of enzymes.

The leaf blade of the Venus flytrap (*Dionaea muscipula*) superficially resembles a steel trap. The upper surface bears six sensitive hairs, any two of which if simultaneously touched cause the leaf to close suddenly, trapping the insect. Digestion and absorption take place, after which the leaf blade resumes its original open position.

None of the above plants are completely dependent on animal prey for nitrogenous compounds, as they are also able to absorb inorganic nitrogen compounds from the soil. Many fungi, however, which are parasites of man and other animals are obligative carnivores and must obtain all their food from their prey.

Pollination by Animals

Unlike animals, plants are unable to move around to find a mate. Thus they have to rely on the movement of animals or the wind to ensure that cross-pollination between different members of the same species takes place. In the tropics few species, other than conifers, grasses, and sedges, are wind pollinated, the majority use insects, birds, or small mammals as pollination vectors. The number of vertebrate pollinators is much higher in the tropics than it is in temperate regions.

Pollination is an excellent example of the mutualism which can exist between plants and animals. For the plant fertilization of ova and production of seeds is of extreme importance and it will devote much energy and food resources to ensure that this is successfully accomplished. The only reason for the production of the often large and brightly coloured tropical flowers is to attract pollinators. Similarly, the powerful scents of many tropical flowers are produced solely to attract pollinators with poor eyesight, or those that are active at night.

However, attraction alone is not a sufficient inducement and will not ensure that an animal visits many members of a particular species, ensuring effective cross-pollination. This can only be achieved if the plant produces special food for the pollinator, in the form of nectar or pollen. The nectar of some flowers contains essential amino acids and is thus highly nutritious, while pollen has a high protein content. Many pollinators, especially insects, live entirely on food produced for them by the flowers of certain species.

Many tropical plants co-evolved with birds and bats. The tropical American humming birds (Trochilidae) are the only pollinators of many species, while the birds can only feed from flowers with long tubes. Thus humming bird plants and humming birds are totally dependent on one another. In the old world, honey eaters (Meliphagidae) and sunbirds (Nectariniidae) pollinate flowers, but the dependence of bird and plant upon each other is not so absolute as with the humming birds. Bird pollinated plants usually have red or yellow flowers set conspicuously on the outside of the plant.

Flowers pollinated by bats are usually white or dull coloured. Bats are colour blind and thus the plant does not need to waste its resources on producing dyes. Instead, such plants synthesize copious nectar and the flowers open at night, when bats are active. In order to force the animal to visit several plants in one night and thus ensure cross-pollination, only a few flowers open on one plant at any one time. Bats range many kilometres each night in search of nectar.

Plants pollinated by insects are variously coloured, blue flowers usually being pollinated by bees and white flowers by moths. Bees cannot see red but they can see ultraviolet light which the human eye cannot. Many bee-pollinated flowers are patterned with chemicals which absorb ultraviolet light. Although invisible to us, these honey guides lead the bee to the nectaries and ensure that it reaches the correct position to deposit pollen from other flowers on the stigma and collect pollen from the anthers.

Many flowers have developed intricate mechanisms to prevent self-pollination, the most common being the staggered ripening of stigma and stamens. The tropical *Catalpa* and *Tecoma* species have stigmas which close after contact with an insect. This helps fertilization to take place and prevents self-pollination. In *Cypripedium* orchids insect visitors enter the inflated lip through an opening at the top but leave by means of openings in the back of the corolla. This ensures that the first contact of the insect is with the stigma and only subsequently does it collect pollen from the anthers.

One great difficulty that plants have had to overcome during their co-evolution with insects and other pollinators is ensuring that the vector visits only one, or at most, a few species at any one time. Otherwise the chances of transferring viable pollen from one plant to another of the same species are poor. Thus the flower colour, shape, scent, and the nutritional qualities of the nectar and pollen have to be aimed at one type of pollinator. Bee flowers, for instance, have short, wide corollas, while humming bird flowers have long throats.

The bee orchids have ensured flower constancy by appealing to the mating instincts of the male insect. The shape and colours of the flower are similar to those of the female bee, while the scents produced by these plants are bee sex pheromones. Thus the male is strongly attracted and attempts to copulate with the flower. In so doing he deposits pollen from another flower on the stigma and collects pollen from the anthers.

The scents of many flowers are similar to insect pheromones, while some insects can use the compounds contained in flower volatile oils to produce their own pheromones. Some plants produce odours which to humans are most unattractive. Such odours have a fishy or amine content, which attracts those insects that feed or lay their eggs on rotting flesh.

Although all plants and their pollinators are dependent on one another to some extent, some have pushed this dependency to the utmost extreme, neither plant nor animal being able to exist alone. This is particularly true amongst the orchids, but other examples also exist. Almost every species of fig (*Ficus*), for example, is associated with a particular species of wasp. The fig needs the wasp to pollinate its flowers, while the wasp needs the fig to nuture its larvae.

In the genus *Yucca* the sole pollinator, the female *Tegeticula* moth, carries a ball of pollen from one flower to another. This she thrusts down the tubular stigma and then lays eggs amongst the ovules she has helped to fertilize. A few seeds always escape being eaten.

Fruit and Seed Dispersal

Another example of mutualism between plants and animals is that shown by the fruit eaters who disperse the seeds of plants. The placing of a seed on the ground some distance from the parent plant is important to the survival of the seedling and of the species as a whole. Seedlings germinating within the vicinity of a vigorous parent are unlikely to develop to maturity. Similarly, many seedlings of the same species germinating at the same time in the same place leads to such intense competition that most, if not all, perish. This is a waste of the parent plant's resources and genetic material. Thus there is intense competition for dispersal agents and the sweet, brightly-coloured fruits produced by many plants are, like flowers, an inducement to animals.

Fruits usually either all ripen at once and are eaten by a variety of seed-dispersal agents, or a few ripen every day over a long period. These are eaten by only one or two species of animal. In the former case the fruits are highly coloured and scented to attract as many dispersal agents as possible. In the latter case the fruits are much less conspicuous, as the specific dispersal agent knows where to find them.

Birds are the most common agents of seed dispersal. Either they spit out the seed before swallowing the berry, or they swallow the whole fruit and the seed passes out undamaged with the faeces. For certain species of plant the passage of seeds through the digestive tract facilitates germination. Some birds, such as toucans, regurgitate seeds after their digestive processes have removed the nutritious fruit.

In the tropics, particularly, many animals other than birds are fruit eaters, while rodents collect seeds and bury them for future use. The seeds of many forest trees are large, so only a few animals can eat them. Bats carry such seeds in their claws and rodents in their mouths, many being dropped. As only one or two seeds are carried away at a time efficient dispersal is ensured even though some seeds are eaten. Unlike temperate plants, few fruits of tropical plants have hooks for attachment to the fur of passing animals.

Because some animals eat seeds as well as fruits many plants have developed poisonous seeds to protect against predation. Such highly toxic seeds are common in the tropics and include those of the Calabar bean (*Physostigma venenosum*), which contains the alkaloid physostigmine, the castor oil bean (*Ricinus communis*), which contians ricin, a toxic protein, and *Strophanthus* species, which contain cardiac glycosides.

Birds and other animals can spread seeds other than by eating the fruits of plants. Water birds carry mud on their feet in which the tiny seeds of water plants are abundant. This explains the rapid spread of such plants to man-made lakes and reservoirs. The plumage of birds can also harbour fungal spores and can be the agent for the spread of plant diseases from one country to another.

The production of ripe fruits with their usually abundant sugar content is very energy consuming for the plant. Thus it needs to ensure that the fruit is only removed when the seed is fully mature. Unripe fruits are usually green and are thus indistinguishable from other parts of the plant. The presence of chlorophyll, responsible for the green colour, is also advantageous as it enables synthesis of sugars to take place. Unripe fruits usually have an unpleasant taste due to their tannin, acid, or resin content. These substances disappear as the fruit ripens and changes colour. Not all fruits are brightly coloured, as those eaten by bats, for example, remain green or yellowish.

Colour is not the only way in which plants attract dispersal agents, ripe fruits usually also have characteristic, strong odours, due to the release of volatile oils.

Suggestions for Further Reading

Barbier, M. (1979). *Introduction to Chemical Ecology*. Longman.
Bowen, H. J. M. (1979). *Environmental Chemistry of the Elements*. Academic Press.
Edwards, P. J. and Wratten, S. D. (1980). *Ecology of Insect–Plant Interactions*. Arnold.
Faegri, K. and von der Pijl, L. (1966). *Principles of Pollination Ecology*. Pergamon Press.
Harborne, J. B. (1977). *Introduction to Ecological Biochemistry*. Academic Press.
Rosenthal, G. and Janzen, J. H. (Eds) (1979). *Herbivores, their Interaction with Secondary Plant Metabolites*. Academic Press.
Vickery, M. L. and Vickery, B. (1981). *Secondary Plant Metabolism*. Macmillan.

Chapter 9

Plants and Man

Introduction

When primitive man scattered the first seeds, cut down the first tree or started the first fire he began an interference with plant ecology which has continued to the present time. This interference has escalated over the ages so that today there are few places in the world where the activities of man have not influenced the environment of plants. Although in some areas man and nature work together to the benefit of both, in many the activities of man have proved disastrous to natural plant life.

The tropical rain forest is a fragile ecosystem which is easily destroyed, particularly by the activities of man. If left undisturbed, however, it remains unchanged for thousands of years. Theoretically most of the humid tropics should be covered in primary forest, the fact that they are not is due as much to man as to climatic factors. The rate at which such forest is disappearing is alarmingly high and unless steps are taken to preserve that which remains this ecosystem will eventually disappear forever from the earth, taking with it the great diversity of species found in such an environment. Many countries have already lost all or most of their primary forest, their vegetation cover being the result of man's interference with nature. It is very important that man should be aware of his influence on the natural environment, so that he can cease, or at least moderate, those activities which lead to devastation.

All species of plants and animals are unique genetic sources which cannot be replaced. Every time a plant or animal becomes extinct the total genetic pool is reduced and valuable characteristics are lost forever. Realization of the necessity to conserve plant genetic material has led to the setting up of gene banks in various parts of the world, such as the International Institute of Tropical Agriculture's gene bank at Ibadan, Nigeria. Nature reserves and botanical gardens also help to preserve species which are in danger of extinction.

The two main influences of man on the ecosystems of the tropics are farming and timber felling. The clearing of large tracts of land for the development of towns and cities has only, so far, occurred on a small scale and cannot be compared with the devastation wrought in temperate regions. Although some localized areas, especially in Zaïre and Zambia, have suffered from the effects of mining activities, these too are insignificant when the tropics as a whole are reviewed. However, mining activities and drilling for oil could become important as tropical countries become more developed.

Agriculture and timber felling are ancient activities of man, but it is only in the last 100 years or so that, due to explosive population increases, they have had permanent and devastating effects on the tropical environment. Large plantations of single species crops, such as rubber, sugar cane, tea, and coffee, etc. have also resulted in the destruction of many natural ecosystems. The problems of land use in the tropics are considerable and must be resolved if vast tracts are not to become useless wastelands.

Fire

Although natural fires caused by lightning, volcanoes, meteorites, etc. must have interfered with tropical vegetation through all stages of its evolution, it is only since man learnt to use fire that permanent changes in many tropical ecosystems have occurred. The main influence has been through the method of cultivation known as slash and burn or shifting cultivation, described below.

Those parts of the tropics which have several months of dry weather are very vulnerable to fire, both natural and man-made. Such areas are also those of the grassland savannas. With their growing points underground, grasses are amongst the most fire resistant of all plants. In fact it is now accepted that a quick, light burn is beneficial to pasture, stimulating the growth of new shoots and increasing their nutritional value.

Whether or not the savannas of the tropics are a direct result of fire cannot be stated with certainty, as such areas also have climates more suited to grassland than to forest. However, fire was and is undoubtedly a factor in the conversion of forests to grasslands. Recurring fire leads eventually to the extinction of non-fire resistant species, thus completely altering the character of an area. Some plant species, however, are so adapted to fire that their seeds are unable to germinate unless exposed to the high temperatures of fire. Thus seeds of the South African proteas and the *Acacia* species of the savannas need the stimulation of fire to germinate.

The natural balance of the soil can be considerably altered by fire, which removes the humus in the surface layer of the topsoil, destroys microorganisms, and increases the concentration of soluble salts. The ash left after a fire temporarily increases fertility, as the minerals once locked up in the living biomass are returned to the soil. The concentrations of phosphorus, potassium, magnesium, and calcium are increased but the highly soluble nitrogen salts are either volatilized or soon leached out. The removal of plant cover by fire can also result in severe erosion of the soil, especially on slopes.

Three types of fire occur – ground fire, surface fire, and crown fire. Ground fire is the most destructive, as it can burn underground for days or weeks, destroying roots, tubers, and seeds. Few plants can survive ground fire, but as a deep layer of humus is necessary to sustain such a fire, ground fire is rare in the tropics.

Surface fire is the most common type of fire occurring in the tropics (Plate 9.1), especially in the savannas. Such fire destroys vegetation at surface level but does not penetrate more than a few centimetres below ground level. Thus only surface roots are damaged. Plants that can regenerate from rootstocks, such as grasses and some shrubs, and plants with bulbs or tubers are able to survive repeated surface fires. Some trees, too, have bark which is thick enough to protect the living tissues from fire. Palms are particularly fire-resistant plants.

Crown fires are very destructive forest fires in which the flames spread from one tree crown to another. Such fires destroy everything above ground level, causing complete devastation. They are also likely to be the cause of ground

Plate 9.1
Surface fires are common in the tropics. (Reproduced by permission of A.A.A. photo, Paris.)

fires. In the humid tropics crown fires are rare, as the moisture in the air and on the leaf surfaces prevents such a fire from starting.

Cultivation

In the tropics cultivation has taken place for many thousands of years and thus it is difficult to determine the original effect of man's interference with natural vegetation. However, in the past such effects would have been little more than local, due to small populations and lack of sophisticated tools and machines. It would seem that a balance between man and nature was achieved, which remained more or less undisturbed until the last century. During the past 100 years or so, however, the activities of man have led to the destruction of much tropical vegetation. Increases in the population of tropical areas have resulted in more land being taken into cultivation or used for grazing animals. The colonization of much of the tropics by Europeans has also had devastating effects on the natural vegetation, as large areas have been cleared in order that plantations of rubber, oil palm, sugar, cotton, tobacco, tea, and coffee, etc. may be grown.

Slash and burn cultivation

Slash and burn or shifting cultivation (Plate 9.2) is practised throughout the tropics and has been the method of cultivation for thousands of years. An area of land sufficient to grow food for the whole family or village is cleared of scrub and smaller trees. Large trees and palms are often left untouched to provide shade, palm wine, or for religious reasons. The cut area is then burnt, dug and planted.

The ash resulting from the burnt debris provides sufficient nutrients for 2–5 years of cultivation, after which the land becomes unproductive and is then left fallow while another cleared area is cultivated.

In the past, when villages were well scattered and their populations small, the land was allowed to lie fallow for 10–15 years. During this time secondary forest became established. Today, however, there is in many regions insufficient land to support the increased populations, and fallow times have had to be considerably reduced. This prevents the establishment of secondary forest and with more frequent fires, non-fire resistant species are fast disappearing. Because the vegetation has less time to build nutrient reserves the ash contains lower concentrations of minerals and thus the fertility of the land is reduced. Eventually hardy grasses of little nutritional value to livestock dominate such impoverished soils. For example, large tracts of land in Asia have become covered with *Imperata* grass, as a result of overcultivation. Although drought resistant and able to withstand fire, *Imperata* has little nutritional value for livestock.

Frequent removal of natural vegetation also alters the microclimate and leads eventually to soil erosion.

Plate 9.2
Slash and burn cultivation. (Reproduced by permission of Paul Almasy, Neuilly (France).)

Although slash and burn cultivation is predominant in the tropics, more permanent types also occur. The rice paddy fields of Asia cover much of the lower ground which is seasonally flooded or can be irrigated. Flood waters continually renew fertility of the soil and thus there is no need for the land to lie fallow. The natural vegetation of these regions disappeared long ago.

Introduction of exotic species

The domestication of plants began when man ceased to be a hunter–gatherer and started farming the land. Thus some crop plants, such as cereals and legumes, are extremely old and in several instances their wild ancestors are unknown.

Man has been responsible for the spread of many plant species throughout the world. Maize, rice, sorghum, sugar, tobacco, bananas, citrus fruits, pineapples, coconuts, and rubber have been introduced to all tropical regions by man. The most widespread crop plants are listed in Table 9.1, together with their place of origin. Ornamental plants, too, have been cultivated by man in places far from their origin. Bougainvillea and hibiscus are now ubiquitous tropical and semi-tropical plants and even temperate roses can be found in gardens in many parts of the tropics.

The introduction of grasses such as *Panicum* species and *Hyparrhenia rufa* to South America from Africa has resulted in the colonization of large areas. Such grasses have proved valuable introductions as they have a greater nutritional value to livestock than the native grasses they replaced.

Man has also been responsible for the spread of many plants collectively known as weeds. Under natural conditions these plants face sufficient competition for their numbers to be kept in check. In man-made habitats, however, such as cultivated fields and gardens, they face less competition and are able to thrive. Weed seeds are easily spread throughout the world. While most are unable to survive in their new habitats, some quickly adapt and even take over from native plants.

Production of new species and varieties

Man has not only deliberately produced new species and varieties of plants but he has also assisted in natural evolution by altering the environment. In a stable environment evolution is very slow as there is little need for change, but a changing environment results in adaptation and thus the evolution of new species. However, man's interference with the habitat of plants has also resulted in the extinction of those species unable to adapt.

The deliberate selection of seeds of first wild and then cultivated species over a long time span has resulted in the production of varieties of the species known as *cultivars* or *cultigens*. All edible cereals grown today are cultivars and in many cases their wild ancestors are unknown. Cultivated members of the Leguminosae family (peas and beans) have a similar history. Most cultivars are unable to survive without the help of man. Grain crops, for instance, have lost their ability to shatter and thus to spread their seeds around – they need man as a seed dispersal agent. Maize, in particular, cannot germinate without the help of man as the cob remains intact until it rots away. Edible bananas have been selected over the years from varieties with no seeds to spoil the texture of the fruit. Thus cultivated bananas are sterile and can only be propagated vegetively.

Cultivars used as food for man and domestic animals are selected for their taste, which has resulted in the loss of bitter flavoured compounds, such as alkaloids, tannins, etc. Such compounds were synthesized by wild plants as a protective device. Slugs and snails when placed amongst clover plants will preferentially eat those varieties which do not contain bitter-tasting cyanogenic glycosides. Thus man has to protect his cultivars against pests and diseases by spraying with toxic chemicals which replace natural defence mechanisms.

Table 9.1
Some important tropical crop plants and their places of origin

Crop	Common name	Botanical name	Place of origin
Cereals	Maize	*Zea mays*	Tropical America
	Rice	*Oryza sativa*	Asia
	Sorghum	*Sorghum bicolor*	Africa
	Millet	*Eleusine coracana*	Africa
	Millet	*Pennisetum typhoides*	Africa
Tubers	Yams	*Dioscorea* spp.	Tropical America, Africa, Asia
	Cassava, Manioc, Tapioca	*Manihot esculenta*	Tropical America
	Sweet potato	*Ipomoea batatas*	Tropical America
Legumes	Lima bean	*Phaseolus lunatus*	Tropical America
	Soybean	*Glycine max*	Asia
	Peanut, Groundnut	*Arachis hypogeae*	Tropical America
Fruits	Bananas	*Musa*	Asia
	Pineapples	*Ananas comosus*	South America
	Mangoes	*Mangifera indica*	Asia
	Papaya, Pawpaw	*Carica papaya*	Tropical America
	Oranges	*Citrus sinensis*	Asia
	Grapefruit	*Citrus paradisi*	West Indies
	Limes	*Citrus aurantifolia*	Asia
	Passion fruit	*Passiflora edulis*	South America
	Avocado pear	*Persea americana*	South America
	Breadfruit	*Artocarpus communis*	Pacific Islands
Vegetables	Aubergines	*Solanum melongena*	Asia
	Sweet peppers	*Capsicum annuum*	Tropical America
	Chillies	*Capsicum* spp.	Tropical America
	Tomatoes	*Lycopersicum esculentum*	South America
Beverages	Tea	*Camellia sinensis*	Asia
	Coffee	*Coffea* spp.	Africa
	Cocoa	*Theobroma cacao*	Tropical America
Spices	Cloves	*Eugenia caryophyllata*	Indonesia
	Vanilla	*Vanilla fragrans*	Tropical America
	Cinnamon	*Cinnamomum zeylanicum*	Asia
	Ginger	*Zingiber officinale*	Asia
Miscellaneous			
	Sugar	*Saccharum officinarum*	South Pacific
	Rubber	*Hevea brasiliensis*	South America
	Sisal	*Agave sisalana*	Tropical America
	Cotton	*Gossypium* spp.	Africa
	Tobacco	*Nicotiana tabacum*	Tropical America

Cultivated plants, too, have often lost other forms of protection, such as hairs, prickles, and tough cuticles, all of which make the plant unpalatable to man and other animals. In the wild, cultivars with no protective devices soon die out.

Since the genetic theory was discovered plants have been bred for specific purposes. The aim of breeders of crop plants today is to produce varieties which will give the greatest yields or best quality under a given set of conditions. Thus plants are bred for drought and/or disease resistance. One of the aims of plant breeders in the tropics is to increase the essential amino acid content of the seeds of cereals, and thus end protein deficiency, which is so prevalent amongst tropical peoples, especially children. Some success has already been achieved with maize and rice. However, the conditions under which many tropical crops are grown are so bad that it is difficult to improve varieties which have been cultivated for thousands of years. Such plants are well adapted to drought, poor soil conditions and local pests and diseases, although yields are often poor. Attempts to improve yields, however, usually result in the loss of drought or pest resistance, so that such varieties can only be grown by those farmers able to afford pesticides or to irrigate their land. The majority of the peasantry of tropical regions is unable to do this.

Timber

The deliberate cultivation of trees for timber is a very recent industry in the tropics. In the past it was always the natural forests which were exploited. When populations were low and there was no export of timber the removal of trees for building and fuel did little harm. However, the export of timber such as mahogany (*Swietenia* spp.) and teak (*Tectona grandis*) has decimated forests of these species. As each tree can take up to 150 years to reach an economic size replacement is a lengthy business.

Removal of teak, mahogany, and other timber trees has not only altered the species composition of primary forests but has also led to impoverishment, as much of the fertility of a forest is locked up in its living members. Removal of large numbers of trees without replanting leads eventually to species-poor vegetation and soil erosion. Particularly bad are the recently planted coniferous forests in some parts of the tropics. These will eventually remove all nutrients from the soil, but return nothing. When the trees are cut down the land will be useless for further cultivation.

In order to conserve forest resources and to put the timber industry on a proper commercial basis, very careful management of remaining forests must be applied. Every tree removed should be replaced, preferably with one of the same species. In particular, hardwoods should not be replaced with softwoods.

Erosion

The rapid rate at which soil is being eroded in the tropics is the direct result of increases in human populations. Although the activities of some large animals, particularly elephants, have led to local erosion of soil in the past, no natural causes have had the devastating effects of man over the last century.

The necessity to grow food for an increasing number of people has led to a reduction in the fallow times of abandoned land. Secondary forest cannot become established before the land is cleared again for cultivation. This results eventually in such poor vegetation cover that the vulnerable topsoil is eroded by wind and rain. In Australia thousands of tons of topsoil have been washed into the sea due to the removal of vegetation cover.

Increases in human populations have also led to increases in the numbers of domestic animals. In the past, overgrazing of pasture was avoided by the regular movement of cattle and goats over large areas of land. Although the nomadic life of herders still occurs to some extent, many have now settled in one place and are grazing their animals over much smaller areas. This, together with the increase in the numbers of animals, has denuded grasslands and led to soil erosion. The effects of overcultivation and overgrazing are particularly obvious in East Africa and South America, while it has been estimated that Mexico has lost almost half of its topsoil through such practices.

The intensive cultivation of one-plant stands by Europeans in some parts of the tropics has also led to the impoverishment of the land and subsequent soil erosion.

The impoverishment and erosion of soil are probably two of the worst problems which have to be faced by the governments of tropical countries. Unless people can be educated in the proper use of land large areas of the tropics will be turned into vast dust bowls, virtually devoid of plants and animals.

Water

Man's interference with the water systems of the tropics has been both deliberate and unintentional. The water balance of an ecosystem depends on the vegetation cover. Removal or alteration of this cover changes the water balance and can result in a higher or lower water table, leading to seasonal flooding or drought. Thus deforestation can lead to flooding or the removal of vegetation turn a dry area into a desert.

Deliberate interference with tropical rivers through the building of dams and artificial lakes has caused considerable change in the ecology of the surrounding regions. Because most lakes can absorb all excess water, seasonal flooding downriver is halted. At first sight this might seem to be an advantage, as the land can then be used throughout the year. However, no flooding results in no renewal of fertility and the extinction of species adapted to periodic submergence. Herbivores, both wild and domestic, which once only grazed the land for part of the year, now forage continuously, causing serious overgrazing problems.

The building of the Kariba dam on the Zambezi River has led to the formation of one of the largest man-made lakes in the world. One advantageous result for the locals has been the increase in the population of *Tilapia* fish, but this has been offset by the simultaneous increase in *Schistosoma* snails, the species which causes the parasitic disease bilharzia in man. These snails prefer slow-moving waters of a lake to those of a fast-running river. Slow-moving water encourages the growth of water plants at the edges of the lake and these provide an excellent habitat for the snails.

Many aquatic plants also prefer slow-moving water and the Kariba lake has been invaded by *Salvinia auriculata*, a water-weed from South America. This has formed floating islands on which other plant species are able to grow.

In order to grow crops on marginal land in many parts of the tropics irrigation schemes have been set up. Unless properly managed, such schemes result in a soil becoming so saline that it is useless for cultivation, as has happened in parts of the Indian sub-continent. Irrigation can increase the salinity of a soil by washing salts down from higher areas or by raising the water table so high that salts originally too deep down to be reached by plant roots are moved upwards by capillarity. Thus irrigated soils must be well drained and receive sufficient water to leach out the salts completely. However, the use of excess water to wash out the salts can cause the rivers below the irrigated land to become so saline that the extinction of many plants and animals occurs. Several irrigation schemes have failed because the importance of these factors was not realised.

Reclamation schemes aimed at making waterlogged soils suitable for cultivation have also failed, due to the high acidity of the reclaimed soils. The lack of air in the soils of marshes and other wet areas leads to a high concentration of hydrogen sulphide, produced by anaerobic organisms. When the soil is drained and aerated this sulphide is oxidized to sulphuric acid which is a plant toxin.

Pollution

Pollution problems in the tropics are not yet as serious as in some temperate regions, although in heavily populated areas they are fast becoming so. Probably the most widespread pollution problem in tropical countries is the dumping of human waste in rivers and streams. This not only upsets the ecology of the rivers by altering the species composition of plants and animals, but also leads to epidemics of chlorea and typhoid amongst people using untreated water for drinking. The nitrogen content of water containing human waste is greatly increased, resulting in the extinction of some species and invasion by others. In particular, there is a large increase in the numbers of algae, which results in a considerble decrease in the amount of light reaching plants and animals below the water surface.

Inorganic fertilizers draining into rivers from cultivated fields have similar effects.

Other causes of polluted water include toxic pesticides and the dumping of residues from mines, paper mills, and sugar processing plants, etc.

Toxic pesticides such as DDT, Aldrin, Dieldrin, etc. also pollute the soil and they are not broken down by soil organisms. As they are passed up the food chain these compounds become concentrated, resulting in an interference with reproduction in some animals and birds. As yet the effects of these compounds on man in unknown. Natural insecticides, such as rotenone and pyrethrin, are broken down by soil organisms and thus do not cause pollution problems. The widespread use of DDT in the tropics to control the *Anopheles* mosquito, which carries the malaria parasite, has resulted in large areas contaminated with the chemical. Unfortunately eradication of the mosquito was not achieved and the latest generations are resistant to DDT. Thus areas once declared free of malaria are now fast becoming infested with the disease again.

Probably the most serious pollution of a tropical area with chemicals has resulted from the use of defoliants in Vietnam. These have caused devastation of vast areas of vegetation.

Atmospheric pollution in the tropics occurs to some extent in large cities, but it is not yet the problem it is in developed countries.

Many of the bedrocks of tropical soils contain ores of precious or useful metals, including gold, copper, titanium, aluminium, and iron. In the past these have been mined by open cast mining which completely removes the soil of the areas mined and thus destroys all vegetation. The dumping of waste material alters the topography of a region and also changes the nature of the vegetation cover.

Suggestions for Further Reading

Baker, H. G. (1964). *Plants and Civilization*. Macmillan.
Bennett, C. F. (1975). *Man and the Earth's Ecosystem*. Wiley.
Ewusie, J. Y. (1980). *Elements of Tropical Ecology*. Heinemann.
Hill, T. A. (1977). *The Biology of Weeds*. Arnold.
Kozlowski, T. T. and Ahlgren, C. E. (Eds) (1979). *Fire and Ecosystems*. Academic Press.
Mellanby, K. (1972). *The Biology of Pollution*. Arnold.
Tivy, J. (1971). *Biogeography*. Oliver and Boyd.
Vickery, M. L. and Vickery, B. (1979). *Plant Products of Tropical Africa*. Macmillan.

Chapter 10

Investigating the Environment

by Dr. John Hall,
University of Dar es Salaam,
Tanzania

Earlier chapters have shown that the environment of plants is complex. The best way to appreciate this is by actually undertaking data collection and interpretation. This chapter, therefore, introduces the methodology of assessing the plant's environment. Attention is confined to a limited range of variables and techniques. Wherever possible, emphasis is on simple and inexpensive apparatus. More sophisticated methods are included only if they are needed for studying factors of special future relevance in resource appraisal.

Standard instruments are not described in any detail: for more information refer to the sales brochures of laboratory or instrument suppliers or, for climate recording instruments, to standard texts and handbooks.

Soil Characteristics

The four soil characteristics which should be determined in most ecological studies are texture, moisture content, organic matter content, and reaction. All these can be evaluated with simple equipment. In addition, because of its importance and rapidly increasing usage, nitrogen fixation rate estimation is outlined.

Soil Texture and Water Content

We are interested in soil texture because this affects the role of soil as the direct source of water for vegetation. However, at any time, the soil storage capacity is unlikely to be either fully utilized, or totally exhausted, and therefore both the soil texture and the water content are determined at the time of sampling. The soil water content of a study area is expressed as a depth of water for a specified depth of soil (usually approximating to a rooting depth).

Textural characteristics determine how much of any water present is actually available for plant growth. The general relationship between soil water-holding capacity and soil texture is known for soils at field capacity (Table 10.1). Once the latter is determined by a textural analysis, the soil storage

Table 10.1
Soil water holding capacity (mm water per metre depth of soil) in relation to soil texture at field capacity

Textural category	Soil water holding capacity		
	Plant-available	Non-plant-available	Total
Coarse sand (<10% clay and silt)	60	40	100
Fine sand (<10% clay and silt)	80	60	140
Loamy sand (10–29% clay and silt)	130	80	210
Sandy loam (30–50% clay and silt)	150	100	250
Sandy clay (45–60% clay and silt)	190	130	320
Clay loam (60–70% clay and silt)	150	230	380
Clay (>50% clay)	210	420	630

capacity can be gauged. If facilities are limited, or an immediate determination is desired, an indication of the soil texture can be obtained from a field test. A soil sample of about 15 ml is taken and moistened sufficiently to hold together without being too sticky to touch. The moistened sample is manipulated in the hand and the feel and pliability determined. Classification is according to the scheme in Table 10.2.

For more precise results a mechanical analysis, carried out in controlled

Table 10.2
Soil texture indications from feel and response to manipulation by hand

Textural category	Feel	Response to manipulation*
Sands	Gritty	No response
Loamy sand	Gritty	Rolling possible, but only into a ball
Sandy loam	Gritty	Rolling into a cylinder possible, bending not possible
Clay loam	Sticky	Rolling into a cylinder possible, bending through only about 90° possible
Sandy clay	Gritty	Rolling into a cylinder possible, bending to make a ring possible
Clay	Sticky	Rolling into a cylinder possible, bending to make a ring possible

* Manipulation involves an attempt to roll the sample into a ball or a cylinder about 15 cm long and attempting to bend any cylinder formed into a ring.

laboratory conditions, is needed. This analysis is conducted on the fine earth fraction of an oven-dried (105°C) sample of soil sieved through a 2 mm mesh. Noting the relative weights of the fine earth and coarser fractions will permit eventual application of findings to the general field situation. About 50 ml of the oven-dried, sieved soil is weighed and mixed with 50 ml of 20 vol hydrogen peroxide in a flask. This is gently heated to ensure destruction of any organic matter present which might invalidate the analysis. When effervescence ceases, heating is stopped and the mixture allowed to cool. Once cool, a dispersing agent such as sodium hexametaphosphate (3 g) or, as a local substitute, washing-up liquid (4 ml), is added and the flask is shaken for several hours to separate individual clay particles. The soil is then washed into a one-litre measuring cylinder and made up to the mark with water. The cylinder is closed and inverted to bring all soil material into suspension and set aside in an upright position. A hydrometer calibrated for mechanical analysis is floated in the cylinder and readings taken at fixed intervals to determine the quantity of solids which remain suspended. The original sample weight less the suspension weight after four minutes is the weight of sand in the sample. The original weight less the suspension weight after two hours is the combined weight of sand and silt, while that remaining in suspension is the weight of clay. The weights of the three fractions are expressed as percentages of the sample weight and the soil texture ascertained by reference to a soil textural triangle (fig. 10.1).

In any soil, two categories of water occur – water available for plant growth and water not available for plant growth. When a soil loses water it is lost first from the available category and only when this is exhausted is non-plant-available water lost. Gravimetric analysis determines the total amount of water remaining, regardless of category. The relative proportions of each category in the soil depend on its texture and from Table 10.1 it is possible to determine the amount of water which is available for plant growth. Periodic resampling of soil by gravimetric analysis is needed to show change in water content over time but only a single sampling is required for textural analysis.

Because repeated sampling from a small area may result in excessive disturbance, *in situ* measurement of soil water content is preferable. This can be achieved relatively cheaply by installing electrical resistance blocks. Such blocks are made of water-absorbing material, usually gypsum or nylon, which comes into equilibrium with the surrounding soil. Two electrodes are embedded in each block and leads from the electrodes are connected to a meter. The meter reading indicates how much water is present in the soil. Standardized blocks and meters are marketed but it should be possible to construct both items from locally available materials and components.

Soil Organic Matter Content

The organic matter content of a soil is important in ecological work because it influences many other soil characteristics. The overall ability of the soil to

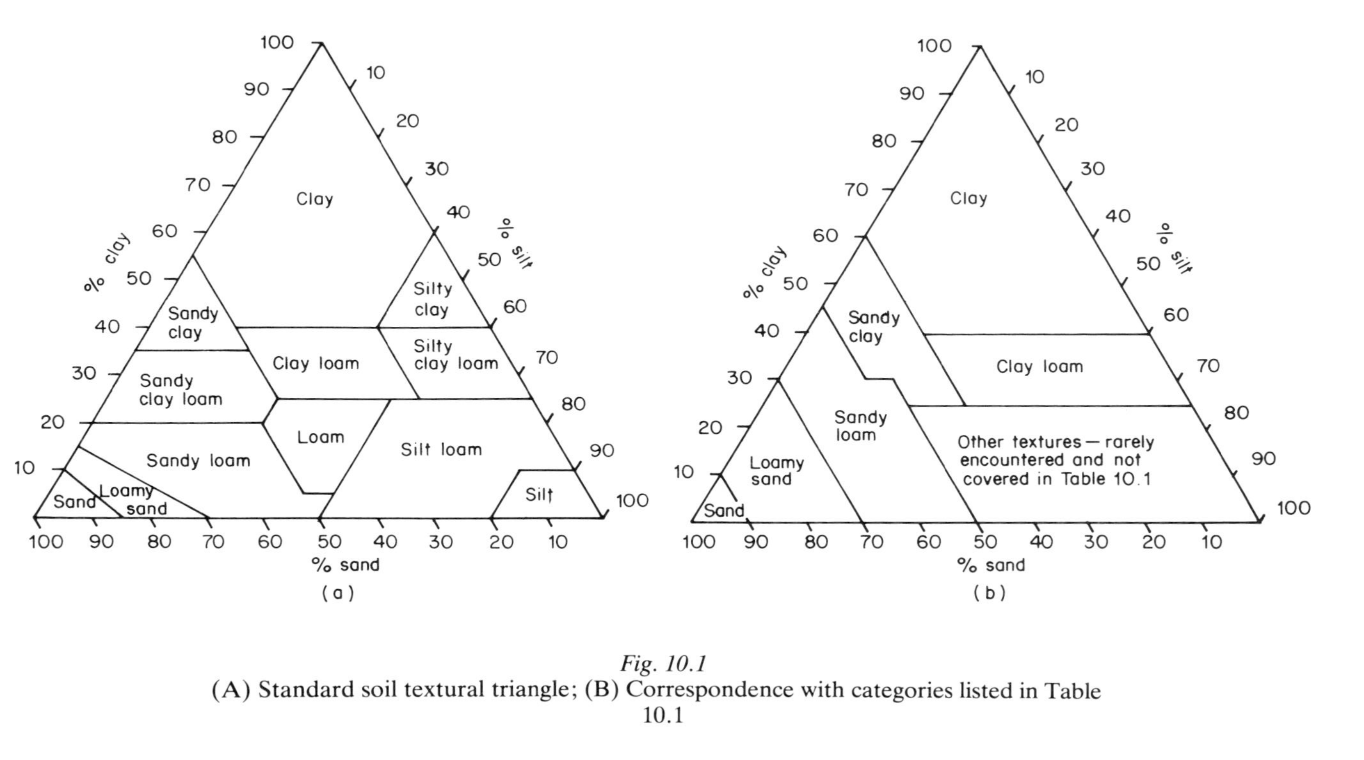

Fig. 10.1
(A) Standard soil textural triangle; (B) Correspondence with categories listed in Table 10.1

support plant growth can be better gauged from the organic matter content than from any other single, easily assessed factor.

Soil organic matter content is simply estimated by determining the weight loss from soil on ignition. A clean crucible is heated strongly for 15 minutes, cooled in the air for 3 minutes and then in a desiccator. When cooled it is weighed. A well-mixed soil sample (2–3 g) is put into the crucible and heated (105°C) in an oven overnight. On removal from the oven, the crucible is cooled in a desiccator and reweighed. By difference, the precise dry weight of the soil sample is obtained. After the second weighing the crucible is heated again – using a bunsen burner – gently for two or three minutes and then more and more strongly. Maximum heat is maintained for 30 minutes, the sample being stirred frequently with a copper wire. After cooling as before, the crucible is weighed for a third time. The loss on ignition is the difference between the last two weighings and is expressed as a percentage by weight of the oven-dry soil.

Soil Reaction

Nutrient availability is strongly influenced by soil pH. When this departs markedly from a neutral value, abnormal growing conditions occur which favour only suitably adapted species. For most crops, the pH ranges within which satisfactory growth can be expected are known; pH is therefore a convenient index of the potential of an area for specific crops. Where relationships between pH and species native to an area are known, these species serve as indicators of sites suitable for particular crops.

Soil reaction is generally measured with pH meters. These are widely available, inexpensive instruments, often battery-powered for field use. Pocket meters are used to obtain a rapid, direct estimation of the pH of any soil. However, when accurate values are needed, a more closely controlled approach is advisable. More accurate instruments are connected to probes inserted into a suspension of soil in distilled water. The suspension is stirred, allowed to come to equilibrium and the meter reading noted. Various soil:distilled water volume ratios (usually 1:1, 1:2, or 1:2.5) are used, the main consideration being to achieve an effective suspension.

Soil Nitrogen Fixation Rates

Natural nitrogen fixation has become an important factor in tropical agriculture due to the rapidly rising cost of commercial nitrogen fertilizers at a time when funds are less readily available. Many plants naturally enhance soil nitrogen status through the atmospheric nitrogen fixation of their associated microorganisms. However, recent studies have revealed considerable variation in nitrogen-fixing capacity, even among those microorganisms hosted by the legumes. Species which fix nitrogen do so at rates which vary with the age of the plant and the prevailing conditions. Studies of nitrogen fixation rates help in understanding the nitrogen economy of many legume-rich ecosystems.

Through such investigations species with the potential for improving mixed crops and agroforestry can be identified.

One widely used technique for estimating nitrogen fixation rates is the acetylene reduction method. The name refers to the use of acetylene as a substrate for reduction by any nitrogen fixing organisms present. The method involves incubating soil or root samples while reduction takes place and measurement of the concentration of ethylene produced in the process.

Possible sources of nitrogen fixing organisms are incubated in flasks of known volume (about 100 ml) fitted with diaphragms. Parallel incubations of control flasks enable detection of contamination or alternative sources of fixation. At the start of incubation, a known volume of air (approximately 10 ml) is withdrawn from each flask via the diaphragm, using a hypodermic syringe and replaced by an identical volume of acetylene. The flask is now positioned in conditions as similar as possible to those normally experienced (e.g. soil samples are incubated in covered pits). The incubation period is critical and thus test runs must be carried out to determine this. Appropriate incubation periods usually lie between ½ hour and 2 hours. If incubation is too brief, the ethylene produced may be in a concentration too low to detect; if it is too long, concentrations which inhibit activity may develop. After incubation, gas samples are extracted for analysis while the test material from the flask is dried (105°C) to contant weight in an oven, so that the source sample dry weight is known.

The gas samples may be stored for a few days in pre-evacuated serum flasks of known volume or entered directly into a gas chromatograph system for immediate analysis. This analysis relies on the identification and measurement of a peak current as the ethylene passes the instrument's detector. Since the dry weight of source material and the incubation time are known, a rate of fixation in grams of nitrogen per gram dry weight of sample per hour can be calculated. If field sampling has been organized to estimate the dry weights on an area basis of material incubated, fixation per hectare over time can be estimated. For long periods this should take account of the times of the year when different climatic conditions prevail. This is why it is important to record the conditions during each incubation and to carry out incubations by both day and night and in cloudy as well as clear conditions. A per hectare figure for annual fixation is especially useful for comparing the potentials of different fixation systems.

Water in the Hydrological Cycle

Water availability for plant growth varies both in space and time and this has a strong influence on the vegetation that develops in a particular area. The measurement of soil water content has already been described. However, while the measurements are a useful guidc to the availability of water for roots, it is important to consider the atmospheric components of the hydrological cycle as these determine the amount of water supplied to the soil (precipitation) and the amount lost from it (evapotranspiration). The determination of such gains and losses together with information on soil texture enables a site water budget to be compiled.

Gains from the atmosphere

In the tropics, rainfall is usually the method by which water enters the soil. Existing meterological stations in national networks are often sufficiently numerous and well distributed to provide a good picture of rainfall patterns. Established networks may prove inadequate, however, when a more precise picture of rainfall gradients in a small but climatically variable area is needed. In mountainous terrain, for example, localized rainfall variations may account for the presence of a wide range of vegetation types. Such rainfall patterns can be discovered from data gathered from a network of rainfall gauges set up through the area. The gauges are located randomly or on a systematic basis, subject to the usual constraints of access and re-finding. Cheap substitutes for standard gauges will suffice and can be purchased or made. If long intervals with appreciable rainfall are likely between readings, the design should provide for this. An example is illustrated (fig. 10.2).

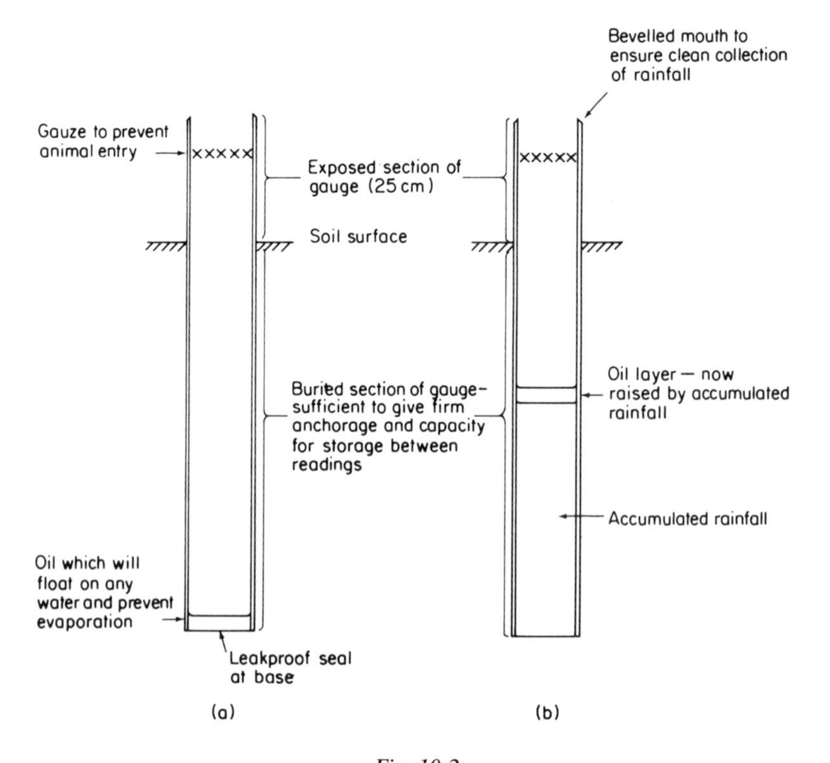

Fig. 10.2
An improved rain gauge made of metal or strong durable plastic (e.g. plastic drain pipe). Readings are taken with a dip-stick. (A) On installation; (B) In use

Losses to the atmosphere

The soil loses water to the atmosphere through evapotranspiration. The extent of evapotranspiration is estimated from meteorological data or observation of evaporation tanks. For water budget preparation (see below) potential evapotranspiration is first estimated and the values are then used in estimating actual evapotranspiration. Several methods of deriving the estimates are available. These differ in the number of meterological data needed. The most favoured method is that of Penman, which makes use of temperature, humidity, wind, and sunshine values as explained in standard texts. Approximations which involve much less calculation and need only values of evaporation from a tank are sometimes substituted. These are, however, less reliable and need to be adjusted for the difference between a water surface and a land surface carrying vegetation, and for the increased rate of evaporation from water in a small tank compared with a larger, more uniform surface. In many parts of the tropics, the most widely installed evaporation tank is of the United States Class A type (fig. 10.3). Recorded values of evaporation from this are reduced to 70 per cent to overcome the effects of small tank size and must be further reduced according to the kind of land surface for which estimates are wanted.

The potential evapotranspiration so determined indicates the demand for water by the atmosphere. The actual evapotranspiration indicates the extent to which this demand can be met. The effects of actual evapotranspiration on any reserve of water in the soil are the basis of the water budgeting procedure. Three water sources may contribute to meeting the evapotranspiration de-

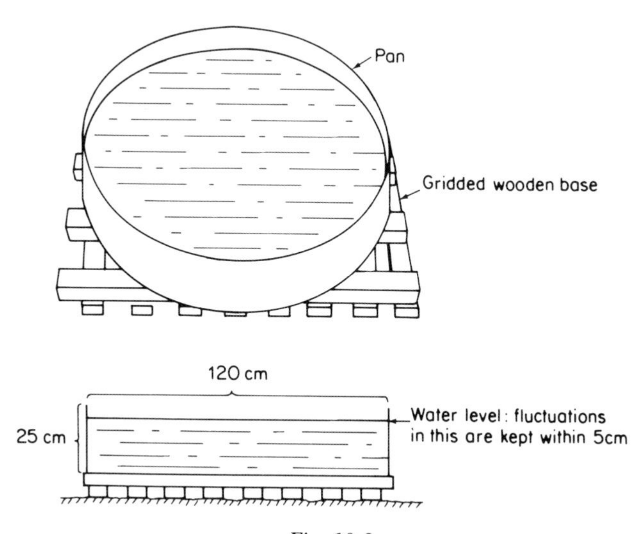

Fig. 10.3
The basic U.S. Weather Bureau Class A Pan for measurement of loss by evaporation from an exposed water surface. Recommended modifications are painting the interior with black bitumous paint to reduce the risk of leakage and addition of a 2.5 cm mesh screen across the top to prevent animals reaching the water

mand – recently received rainfall, water held in the soil and available to plants, and water in the soil but outside the root range. The first two are used by plants until each is exhausted or the potential evapotranspiration demand is met. In deep soils it is assumed that after rainfall and plant-available reserves in the root range have been used up, one-tenth of any outstanding demand can be met by water outside the root range. When the full potential demand is met estimates of actual and potential evapotranspiration are identical; otherwise actual evapotranspiration is lower.

Formulating a water budget

For a given volume of soil of a particular texture, upper and lower limits of possible water storage can be gauged as already described. Water budgets show how the soil water content varies between these limits with time. A time unit (e.g. week, ten days, or month) appropriate for the study is chosen and data are summarized accordingly. A direct measurement of soil water content on a particular occasion (e.g. by gravimetric analysis) is a useful starting point. As an alternative, however, it may be possible to assume from seasonal rainfall trends that on a particular occasion plant-available water stored in maximal (soil at field capacity) or is zero. This approach is most useful when old data are being analyzed.

From the data assembled a balance sheet is constructed. Input is represented by precipitation, output by actual evapotranspiration. The water stored in the soil represents a continually changing balance. When precipitation is in excess of potential evapotranspiration and the soil store is completely filled any further water is lost from the system as surface run-off or deep drainage, eventually reaching rivers. The quantities of surplus water can be estimated. An example of a water budget tabulation is given as Table 10.3. Trends of variation in each budget component can be gauged and expressed diagrammatically, the most generally useful being to show changes in the state of the soil store (fig. 10.4).

Radiation and Temperature; climatic diagrams

The dry matter production of a plant, with an optimal water and nutrient supply, is determined by the radiation climate. Through irrigation and fertilizer applications a measure of control over water and nutrient supply is possible. Evaluations of radiation, and of radiation-dependent factors, thus give a useful idea of the production potential of an area.

Radiation

Meteorological station records will prove of value when a general characterization of a locality is sufficient. The commonest practice is to express incoming radiation as an average of the hours of bright sunshine per day for every month.

Table 10.3

A water budget tabulation – for a sandy clay soil 50 cm deep at Ijaiye Forest Reserve, Nigeria, for 1975

A. Demand and Supply

Period	Demand (PE)	Income (P)	Supply: From P	Supply: From soil: Plant-available in root range	Supply: From soil: From outside root range	Total (AE)
January	139.1*	13.4	13.4	0	12.6	26.0
February	134.5	131.3	131.3	0	0.3	131.6
March	134.8	83.5	83.5	0	5.1	88.6
April	125.5	127.4	125.5	0	0	125.5
May	109.7	169.3	109.7	0	0	109.7
June	86.9	223.2	86.9	0	0	86.9
July	81.1	157.6	81.1	0	0	81.1
August	76.9	51.4	51.4	25.5	0	76.9
September	73.8	176.5	73.8	0	0	73.8
October	102.0	278.8	102.0	0	0	102.0
November	116.5	18.0	18.0	78.0	2.1	98.1
December	130.1	26.9	26.9	0	10.3	37.2

B. Change in Plant-available Water Content

Period	Rainfall surplus (P–PE)	Soil storage situation for plant-available water: Start of period: Water held (S)	Start of period: Unused storage capacity	End of period: Water held (S)	End of period: Unused storage capacity	Change in water stored (ΔS)	Water remaining after full capacity reached (R)
January	no surplus	0	78.0	0	78.0	0	
February	no surplus	0	78.0	0	78.0	0	
March	no surplus	0	78.0	0	78.0	0	
April	1.9	0	78.0	1.9	76.1	+1.9	
May	59.6	1.9	76.1	61.5	16.5	+59.6	
June	136.3	61.5	16.5	78.0	0	+16.5	119.8
July	76.5	78.0	0	78.0	0	0	76.5
August	no surplus	78.0	0	52.5	25.5	−25.5	
September	102.7	52.5	25.5	78.0	0	+25.5	77.2
October	176.8	78.0	0	78.0	0	0	176.8
November	no surplus	78.0	–	78.0	0	−78.0	
December	no surplus	0	78.0	0	78.0	0	

* All figures are mm depth of water.
Soil water content at field capacity: 78 mm plant-available water.
Assumed situation at beginning of January: 0 mm plant-available water.
AE: estimated actual evapotranspiration; P: precipitation; PE: estimated potential evapotranspiration; R: run-off and deep drainage; S: soil store of plant-available water; ΔS: change in soil store of plant-available water.

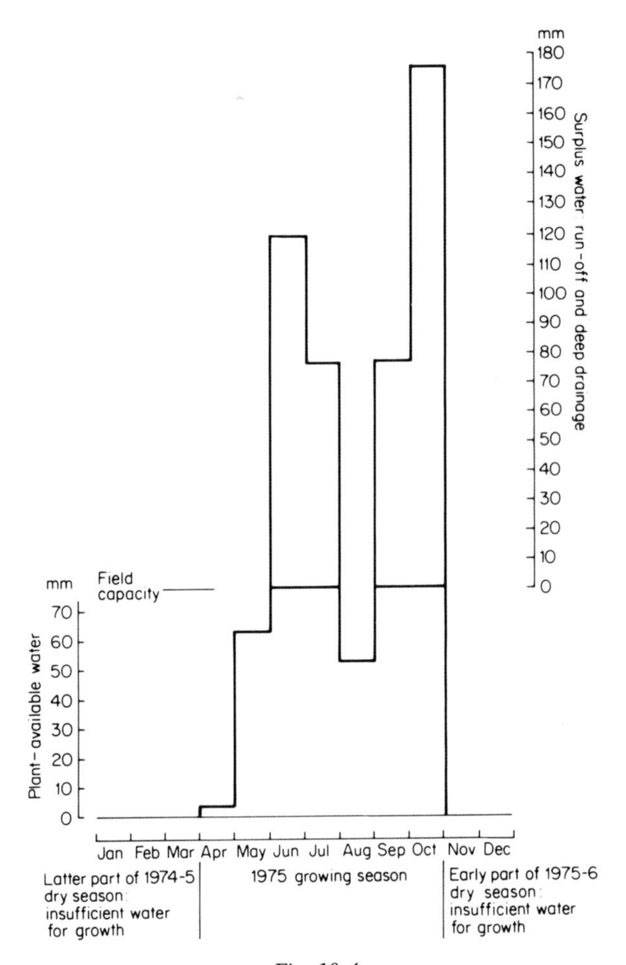

Fig. 10.4
Ijaiye Forest Reserve, Nigeria: diagrammatic representation of water budget for 1975 in sandy clay soil with a rooting depth of 50 cm

These averages can be reported as percentages of the maximum possible bright sunshine as long as the day length is known, and this can be determined from standard tables. Percentages below 100 reveal the influence of clouds and atmospheric dust. Equations incorporating constants have been developed which allow sunshine duration data to be converted to net energy income. Nowadays, however, advantage is usually taken of net radiometer data collected at major meteorological stations.

Net radiometers incorporating thermopiles and solarimeters are appropriate for more thorough reviews of the overall radiation climate. Such instruments, however, need careful installation, calibration, and maintenance. For most ecological work it is enough to limit investigation to two aspects of the radiation climate: light intensity (assumed to correspond to the incoming radiation in the wavelengths associated with photosynthesis) and temperature.

Light intensity

In ecology light intensity is usually studied to see if the degree of shading influences vegetation type and development. Most of the cheaper equipment used to assess incoming radiation is not readily calibrated in absolute terms. Emphasis is therefore on relative values at different points in time or between instruments in different locations at the same time. A simple light meter, such as the traditional photographic light meter with a sensitive selenium cell, is suitable. The cell should be mounted separately from the meter and screened with a filter to admit only light of wavelength 400–700 nm. In use, the cell is positioned horizontally, facing down to receive light reflected directly upward from a white paper surface 50 cm below. Suitable shielding restricts light received by the cell to that reflected off the paper. The irradiance can be expressed in lumens per square metre if the instrument is suitably calibrated but a satisfactory expression as an energy flux is difficult because of variability in light quality and the quantum energy of light of different wavelengths. In practice, variations in light intensity in all three spatial dimensions and in time, in complex vegetation, present considerable difficulties in any such studies. The reader is referred to published accounts for details of ways in which this problem has been approached.

Temperature

Mean annual values, and seasonal and diurnal trends, are the aspects of the temperature regime of greatest ecological interest. Mean annual values, for example, have commonly been adopted as one of the criteria for delimiting ecological and climatic regions.

Where an adequate network of meteorological stations exists, data on temperatures will be readily available. However, in areas with a poor coverage, advantage can be taken of the constancy of soil temperatures. At a depth of about one metre, soil temperatures approximate to the mean annual air temperature. A hole is bored to this depth with an auger and the temperature measured with a thermometer inserted in the hole and allowed to stabilize. The boring should not reach the water table and some replication is desirable. This is an especially useful technique for assessing temperature conditions in remote mountainous terrain in a single visit. Observations may be extended to reveal lapse rates if the operation is repeated at several different known elevations and will allow 'mass heating' effects to be detected; large land masses retain more heat and tend to give rise to warmer climates than small masses of the same elevation.

Ambient temperature exerts a marked effect on the rates of biological processes. In addition to mean values, the extent of departures from mean conditions are therefore of interest. Thermometers screened from direct isolation are used. These are read at predetermined times. If thermometers are of the maximum/minimum type, the extremes of a known preceding period (usually 24 hours) are also indicated. Thermographs which record continuously on a chart for weekly or longer periods are also available and these may be combined with a humidity recorder. The advantage of such automatic instruments is obvious but they are costly – especially if use in the field is planned – and need careful and regular resetting to retain accuracy. To allow necessary corrections to the records to be made, a maximum/minimum thermometer is used as a standard with the thermograph.

If the temperature at different places within a small area is to be monitored, a simple, portable instrument such as a whirling hygrometer is adequate. This can also be used to measure humidity or vapour pressure with the help of tables or charts supplied with the instrument. Since the instrument is ventilated when used, no screen for protection against direct insolation is needed. When greater distances separate observation points more expensive automatic recorders can be installed, or arrangements made for simple instruments in each place to be read by different observers.

Climatic diagrams

Often it is unnecessary to measure climatic factors in a study because interest centres on the general features of the climate and good records are kept by local meteorological stations. However, the station's official records will not always be summarized in a form particularly useful for ecological studies. For ecological purposes, the 'climatic diagram' is widely regarded as a convenient summary of conditions. Raw data on rainfall and temperature are combined to show seasonal trends in their values and how they interact. Rainfall values are depicted as monthly totals and temperatures as mean monthly values (the average over the month of the mean of the daily temperature maxima and minima). An established convention is followed in drawing the diagram. Time is represented on the horizontal axis and temperature and rainfall on the vertical axis. For rainfall below 100 mm a rainfall scale interval of 20 mm equates to a temperature interval of 10°C and a time interval of two months. The temperature scale is marked only to 50°C as no higher value is needed. To save space the rainfall scale is modified so that above 100 mm calibrations are for every 200 mm (fig. 10.5). Total annual rainfall and mean annual temperature are indicated at the top. Shading is also in accordance with convention (see fig. 10.5). Following the conventional layout and shading enables rapid comparison of diagrams produced for other data sets even if reproduced at different scales.

The number of months when over 100 mm rainfall is received are important, as this indicates the extent of leaching conditions. The duration of the period when the rainfall line runs above the temperature line shows the length of the rainy season, itself roughly equivalent to the growing season, while the relationship between periods of high rainfall and periods of high temperature indicates whether the area enjoys summer or winter rainfall. Use of these diagrams to estimate growing season length is, of course, less satisfactory than

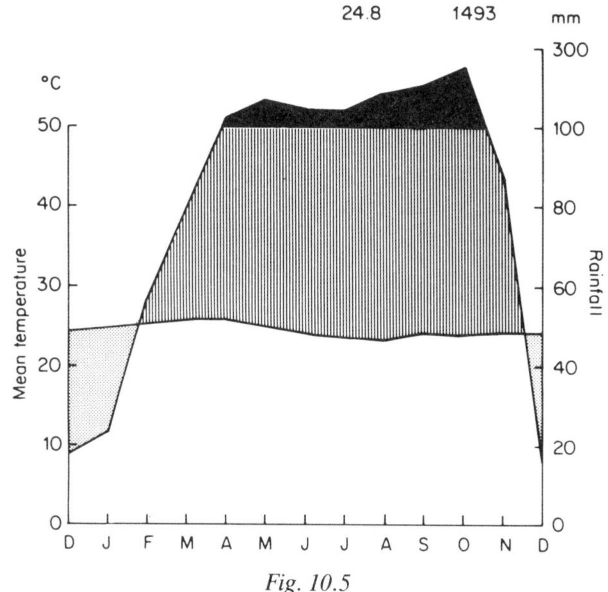

Fig. 10.5
Climatic diagram for Berberati, Central African Republic

procedures which take account of soil water. It is, nevertheless, an improvement upon estimates based on rainfall only, and the speed with which the diagrams can be prepared, if the data are immediately available, is a great advantage.

For a series of years, plots can be made of a complete sequence of individual months (which shows clearly any year-to-year variation) or of average values for each month over several years, giving a more general picture. The latter is the more common practice and is usually sufficient for areas of predictable climatic pattern.

Atmospheric Measurements

Carbon dioxide concentration

Carbon, in the form of carbohydrates and lipids, is the means by which plants store and transfer energy. The source of this carbon is the carbon dioxide in the atmosphere and estimation of the concentration of this gas is a fundamental step in many studies of plant production. Carbon dioxide in the microclimates of plants fluctuates according to diurnal and seasonal variations. Such fluctuations, together with estimates of carbon dioxide exchange under conditions of known temperature and humidity, can be determined using a gas chromatograph. The importance of this technique is that it does not require the destruction of any plant part. Air is drawn alternately through a chamber enclosing leaves *in situ* on a plant and from a nearby source where there is no contact with a plant. The result is a fluctuating meter or chart reading showing how the leaves influence the air passing them. Precautions are taken to ensure that temperature and humidity values for the air sampled are not modified by the sampling equipment and that the flow rate through the chamber is constant. With suitable instrumentation incorporated in the system, transpiration rates as well as carbon dioxide exchange rates can be determined. If relationships between leaf dry weight and surface area have been established previously, the rates measured can be expressed in terms of plant dry weight or surface area.

Wind direction and velocity

Details of prevailing wind conditions are important when considering evapotranspiration processes, attempting to correlate seed and pollen dispersal with environment, and for determining degrees of site exposure. Data on wind direction and velocity extracted from past meteorological records are sufficient if interest is limited to diurnal and seasonal patterns in prevailing winds. When information is needed about wind conditions within vegetation, however, separate determinations must be made. Ideally, instruments connected to automatic recorders would be used, although they are costly. Often there are two parts to the instrument – a set of cups attached to a rotating frame (cup anemometer) and a vane.

The cup anemometer indicates windspeed. A current is generated as rotation takes place or a counter is activated. The speed is deduced from the current generated or the rate of increase of the counter. This instrument responds only to movement in the horizontal plane and is fairly bulky; when general air movement or turbulence is of interest, or space is very limited, other devices such as hot-wire anemometers can be used.

The vane indicates wind direction. Again the instrument operates only in response to movement in a horizontal plane. Thus on installation it needs careful levelling. In use, the wind direction determines the power of a current flowing to a recorder marking a suitably calibrated chart. Where expenditure has to be limited readings must be made at intervals rather than recorded continuously.

Fauna and the individual plant species

Plant–animal interrelationships have long been known. For many plants, particular animals are agents of pollination and dispersal. Other animals, however, which feed on the plant, play a negative role. The status of a plant species within a community cannot be properly appreciated unless some consideration is given to its relations with the animals present.

Pollination

The simplest assessments of animal involvement in pollination entail careful observations of flowers and their visitors. During a flower observation programme each visitor to the plant selected is listed and the times and duration of the visit recorded. Also any distinctive behaviour on the part of the visitor is noted. Knowledge of flowering behaviour will help in planning observations. Some flowers are open only at certain times of the day while others release scent to attract pollinators according to a diurnal pattern.

Not all visits are by pollinators; some visits are specialized nectar thieves gaining access to the flower without passing the anthers or are too small to brush against them. To distinguish these from the genuine pollinators it is necessary to trap them as they leave the flower and examine them for signs of pollen. Pollen from the animals is compared with pollen taken directly from the flower. Parallel observations on other species flowering at the same time are useful as they identify those animals visiting only one kind of flower and thus further restricts the list of possible pollinators.

Cages designed to keep out animals can be used in experiments to see if self-pollination is possible. Insects should be allowed to enter some cages but not others and the cages should contain plants with stamens removed before anthesis, plants intentionally 'selfed', and plants left intact and undisturbed. Visitors to accessible plants should be recorded. It is important that the plants remain in the cages until seed has developed. The success of pollination is then assessed and the effects of the various treatments are gauged by statistical comparisons.

More refined studies of plant/pollinator relationships may involve appraisal of the efficiency and range of the insect as a pollen carrier. Newly emerged insects free of pollen and marked to allow re-identification during observations or on recovery from a nest or hive are used. Fine oil paints or rapid-drying cellulose paints are suitable for marking. Sightings of the marked insect can be recorded. On recapture, the pollen carried can be removed by washing the insect in 70 per cent alcohol, identified, and assessed quantitatively.

From the point of view of gene exchange, interest is in the range over which pollen may be dispersed from a plant. An aqueous suspension of a pigment, such as the fluorescent Helecon 2267, is applied to the stamens of an open flower with a small brush. Trapped insects and flowers that were not specially marked can later be examined under ultra-violet light. If pollen from the marked flower is deposited on the flowers tested or carried on the insects, it will fluoresce. Results not only indicate the type of pollinating agents used by the plant but also allow a pollen dispersal shadow to be constructed.

Seed dispersal

The most easily studied aspect of seed dispersal by animals is identification of the animal species involved. Direct observations of fruit-eating animals are made and these are routine in bird and large mammal field studies. Complementary observations are needed, however, to show whether the animals spread viable seed to places where it can germinate. Association of plant occurrence with animal trails provides useful indirect evidence. Germinating seed present in dung samples of known animals is easily undertaken and helpful for animals difficult to observe directly. Seed can be extracted from dung and germinated under controlled conditions. If comparison is made with seed from identically treated unconsumed fruit, the effect of passage through the animal can be assessed.

Plant-eating animals

Studies of the association between food plants and animals are of considerable ecological importance. Insect–plant relationships are the most conveniently studied as most plant-eating insects are large enough to be readily detected and small enough for the larval phase to be completed on a single host plant.

At its simplest, the study of plant–insect relationships involves listing the insects associated with a particular plant species. Lists prepared at various times of the year reveal seasonal changes in the range of consumers, while parallel observations on other species help in separating specific from non-specific feeders. The insects are detected by a careful search or by separating them from the plant. Chemicals such as dilute pyrethrum applied in the form of a mist blown into the foliage will stupefy the insects so that they fall onto a tray or cloth. Insects are usually removed from dense herbaceous vegetation by sweeping it with a strong net.

To study a species in greater detail, programmes are drawn up for monitoring the population dynamics of potential pests or simply to establish the kind of population present at one time or at intervals over a seasonal cycle. In more detailed work, insect populations are determined in absolute terms for each phase of the plant's life cycle. This introduces complications in sampling the insect population and in relating it to the food resource. Firstly, the spatial variation of the plant species must be estimated so that it can be expressed reliably on a unit area basis. Next, the insect population is assessed in each of a series of randomly or systematically located units of fixed ground surface area. All the insects present in each unit are counted after location by searching (eggs and larvae) or after anaesthetizing (larvae) and collecting them as they fall. Adults are trapped in cages. The amount of vegetation searched or screened is recorded to allow the population sample size to be suitably related to the total amount of the plant present. The soil is searched to an appropriate depth in each unit for pupae, while non-subterranean pupae are located by searches. The adults are counted after emergence from pupae carefully transferred to cages.

Where estimation of absolute numbers is impractical, relative assessments of abundance may be made. The proportion of a collected sample that can be attributed to any one species or group of species is determined. Changes with time in these proportions are revealed by repeated samplings. Only limited interpretation of relative values is possible, however, and much more easily gathered qualitative data may be almost as useful.

Fauna and Vegetation

Heterogeneity is a common feature of the vegetation in areas with populations of large wild animals or of that used as rangeland. The animals themselves contribute to the maintenance of this stage. When a management plan is being prepared it is important to recognize that the different types of vegetation in the mixture may respond differently to environmental influences.

Direct observation of the animals concerned will establish the range of plants eaten. Ideally, however, this should be followed by an attempt to assess the impact of the animals on each vegetation type and to estimate the extent to which each is utilized (e.g. grassy types for grazing; woody phases for browsing). If such observations are extended to individual plant species the data collected will be even more valuable.

Observations on feeding behaviour can provide a more detailed picture of utilization intensity if it is assumed that the proportion of feeding observations referable to each type of food corresponds roughly to its importance in the diet. If observations are extended to cover the actual duration of feeding and the proportion of time devoted to each food item the results are further refined. Seasonal variations are taken into account by repeating observations at regular intervals (of one month or less) throughout the year.

When absolute values of the plants consumed are to be estimated exclosures, which keep the animals out, or enclosures, which keep them in, are used to control access to the food under study. Within the areas demarcated the quantities of grazed plants and browsed plants are assessed.

Grazed plants are assessed by clipping, drying, and weighing the sward in replicated samples within exclosures. The weights are expressed on a unit area basis. Comparison with figures from areas open to animal access will reveal the amount of the sward eaten by animals. Seasonal changes in grazing intensity may be due to variations in chemical or water content of the vegetation. These can be monitored by analysing subsamples of the clipped material.

Quantitative assessment of browsed plants is complicated by the diversity of the woody components in the community. Here a more direct evaluation is needed. Marked shoots on browse plants in enclosures are characterized with respect to length, leaf numbers, and leaf areas. The amount of material contributed by the marked shoots is determined as a proportion of that available in the enclosure. After this initial assessment, animals are admitted to some enclosures but remain excluded from the others. Periodic re-assessment can show the trend of utilization and whether this has a seasonal basis. Again, patterns detected may be readily explained if moisture and chemical contents are monitored. Sub-samples of browse, characterized in the same way as the marked shoots, are used to determine the dry weight per unit area.

Provided the composition of the vegetation is assessed on each sampling occasion, preferably in quantitative terms, any sequence of change can be taken as a guide to the status of the land. This represents the long-term impact of the animals upon the vegetation. It is possible to classify range plants as 'decreasers', 'increasers', and 'invaders'. Decreasers are palatable, nutritious species sensitive to heavy grazing. Increasers are less palatable and tend to replace decreasers when grazing is heavy. Invaders are undesirable weeds of low nutritional value; they are little affected by grazing and proliferate reducing the vigour of decreasers and increasers. Comparison of the proportions of the sward contributed by each class, and the associated percentage of bare ground, on successive sampling occasions indicates whether the vegetation is stable in composition or (as a resource) is improving or deteriorating.

Erosion and fire as agencies for habitat change

Over much of the tropics the rapid increase in human populations is interfering with traditional patterns of land use and, increasingly, this interference leads to land degradation. As a result, soil erosion and burning regimes are increasing and the assessment of their effects is essential in formulating effective policies to arrest land deterioration and promote rehabilitation.

Erosion

Catchment observations

A detailed study of erosion includes the monitoring of sediment loss in streams and rivers of a catchment area over time. Losses are related to the surface area and physical characteristics of the catchment, its rainfall, and the rate of water discharge. Specialized equipment is used to measure discharge rates (usually with a metered weir), suspended sediment (with depth-integrating samplers), and bed-load sediment (with stabilized collecting baskets). Such determinations are mostly carried out by the professional hydrologists or water engineers.

Run-off plots

Run-off plots offer a less sophisticated approach for erosion studies. The aim is direct assessment of soil loss within catchments through observations from small units. The units can be replicated, if facilities permit, to enable results to be applied to the catchment as a whole. The data collected refer to sediment transported by surface water flow to a specific collection site from a defined area, isolated inside a low barrier (fig. 10.6) from surface water moving over the surrounding land. Several precautions must be taken in establishing such plots. The barrier must be inserted into the soil to a depth of at least 10 cm and project above it 30 cm. After installation, time for stabilization of the surface must be allowed before data are collected. The lowest point in the plot must be the sediment collection point and the slope should be uniform. Since there may be cumulative effects of run-off down the slope, a minimum length is needed to secure acceptable figures – 20 m has been suggested; if smaller plots are

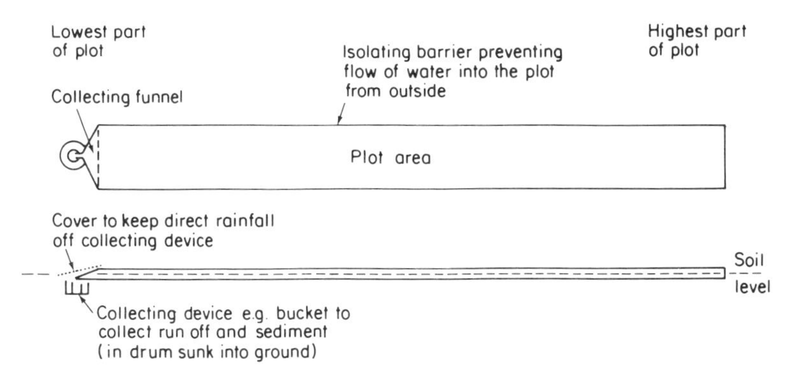

Fig. 10.6
Arrangement of a run-off plot

established emphasis should be on relative amounts of transported sediment and other variables in comparable plots under a range of defined conditions. Devices to funnel the water to the collection point must be shielded from direct entry of rain.

From the results of run-off plot studies, interrelationships of soil losses (g m^{-2}) over a period of time with rainfall, run-off, slope and vegetation can be deduced and nutrient losses can be calculated. The water collected deposits the soil carried as sediment in the collecting receptacle or contains it in suspension. If volumes are small enough, this soil can be filtered on to a filter paper of known dry weight, dried (105°C) and weighed. If a large volume of water has collected, this is subsampled. After initial drying and weighing, the sample is ashed (550°C) and reweighed. Allowance is made for the effect of ashing the filter paper, and the organic fraction (weight lost on ashing) and the mineral fraction (sediment remaining) are recorded separately. Chemical analysis of the ashed material will indicate those nutrients lost during run-off.

Pin and root exposures

Erosion and deposition can both be investigated through observations on specially installed, vertical, calibrated pins, or stakes. These are arranged randomly or in a systematic pattern, and the length exposed is recorded at intervals over a period of time. If enough pins are installed, soil erosion/deposition rates in different units within a site can be monitored for comparative purposes. Average changes between successive recording occasions can be calculated for each unit and, as long as the total area of the unit is known, expressed in volume terms. By making suitable complementary records at the site, erosion can be related to slope, soil type, and the nature of the vegetation. The disadvantage of this method is the lack of provision for relating soil losses to run-off water volumes.

An even simpler method is to quantify existing *signs* of soil loss, such as tree root exposure, where an original surface level for a known past date can be postulated. The vertical difference between past and present surfaces is measured and a mean annual rate of erosion calculated. Replicated and randomized observations are needed to avoid unintentional bias.

Fire

Fire is an extremely variable factor in both intensity and frequency. In some years, the vegetation may escape burning completely while when it does occur fuel could be maximal, allowing the fire to spread extensively. Any intermediate state between these extremes is also possible. The full extent of the effects of fire is determined by monitoring the vegetation during the periods between successive fires. In order to express the results quantitatively the characteristics of the fires themselves have to be determined.

Burning experiments

In most studies of fire emphasis is on vegetation stability and change in experimental plots. Treatments traditionally include freedom from burning, an early (therefore less intense) burning regime and a late (very intense) burning regime. Much information can be gained from observation of such plots. Additionally, a control treatment should be incorporated wherever possible. This should consist of the natural fire regime prevailing in the area. The control plots should burn each year at the same time as the land surrounding the experiment. All experimental treatments should be replicated and those plots totally protected or due to burn only on prescribed occasions should be enclosed by effective fire-breaks – strips 10 m wide maintained free of vegetation. Each experimental plot should be a unit sufficiently large to be comparable in overall structure and composition with every other one at the start of the experimental period. Furthermore, every one should be assessed in detail when the experiment commences, before the first use of fire. Prescribed burning is implemented on fixed dates in regions of low year-to-year climatic variability. In regions of high year-to-year variability a date set in accordance with the establishment of a defined set of circumstances may be more appropriate. Early burning, for example, could be fixed for the day after a specified number of consecutive rain-free days following the end of the main rainy season.

It is advisable to monitor the conditions at the experimental site with respect to rainfall, humidity, and temperature and, within each plot, to follow trends of fuel accumulation and decrease. The conditions needed for permitting a fire of a particular intensity to burn should be known. In particular, care should be taken to record the conditions preceding and prevailing during the burning. With this information it becomes possible to assess the extent to which a prescribed treatment resembles the natural burning regime. If continually recording instruments are not installed, special arrangements should be made to record the wind speed, temperature, and humidity conditions associated with every burn.

Fire temperatures

During burning, it is helpful if the characteristics of the fire itself are assessed. This ideally involves measurement of the temperatures of the fire at specific points – replicated to show the horizontal and vertical gradients created. Several techniques for doing this have been devised. The simplest is using compounds (e.g. temperature-sensitive paints) whose nature visibly and permanently changes if they are exposed to temperatures above a known limit. A group of metal strips, each with a coating responding to a different temperature threshold, is installed at each recording point. After the fire has passed, a check is carried out to see which strips have changed colour and which have not. For this technique the range of temperatures covered should extend from ambient (20°–30°C) to 1000°C.

More precise temperature measurement is possible with a thermocouple, though replication cannot be as extensive as with paints if costs are to remain realistic. An appropriate metal combination for the temperatures likely to be encountered in fire studies is platinum (87 per cent)/rhodium (13 per cent). A particularly useful refinement is to connect any thermocouple to a continuous recorder. This makes it possible to determine the duration for which any temperature reached is maintained. The recorder must occupy a position where it will not be damaged as the fire passes.

Fire history

The fire history of an area is often reflected in the characteristics of its vegetation but is very difficult to document. Many parts of the tropics now benefit, however, from regular satellite photography coverage at intervals of only two or three weeks. Where this is available, it can provide indications of the extent of burning and the fate of a given area can be expressed in terms of the approximate burning time. To complement information from satellite photography, it is constructive to assess the degree of fire-risk at the time of burning. This indicates the intensity of the fire and can be assessed from basic meteorological data for the area in question, subject to assumptions of the fuel available. The climatic data used are daily records of rainfall, air temperature, relative humidity and wind speed. A circular slide meter has been developed which, when set on the basis of these parameters, ranks the fire-risk. Ranking is on a scale of 0–100, corresponding to conditions ranging from those when burning is impossible to circumstances where any fire is likely to be uncontrollable. Early burning in savanna corresponds to indices of about 2–10 and late burning to indices from 5 (areas of sparse vegetation) to 16 or more (savannas of wetter zones). Since, within the tropics, natural vegetation subjected to fires is predominantly grassy, indices exceeding about 25 are unlikely. For plantations of high fire-risk, such as those of conifers established in dry areas and furnishing far greater quantities of fuel, indices may be much higher.

Suggestions for Further Reading

Ahn, P. M. (1970). *West African Soils*. Oxford University Press, London.

Black, C. A. (Ed.) (1965). *Methods of Soil Analysis Part 2. Chemical and Microbiological Properties*. American Society of Agronomy Monograph, **9**.

Evans, G. C. (1956). An area survey method of investigating the distribution of light intensity in woodlands, with particular reference to sunflecks. *Journal of Ecology*, **44**, 391–428.

Evans, G. C., Whitmore, T. C. and Wong, Y. K. (1960). The distribution of light reaching the ground vegetation in a tropical rain forest. *Journal of Ecology*, **48**, 193–204.

Ewing, G. W. (1969). *Instrumental Methods of Chemical Analysis* (third edition). McGraw-Hill, New York.

Fitzpatrick, E. A. (1974). *An Introduction to Soil Science*. Oliver and Boyd, Edinburgh.

Free, J. B. (1970). *Insect Pollination of Crops*. Academic Press, London.

Grimsdell, J. J. R. (1978). *Ecological Monitoring*. African Wildlife Leadership Foundation Handbook, **4**.

Hardy, R., Holsten, R., Jackson, E. and Burns, R. (1968). The acetylene-ethylene assay for N_2 fixation: laboratory and field evaluation. *Plant Physiology*, **43**, 1185–1207.

Hopkins, B. (1965). Observations on savanna burning in the Olokmeji Forest Reserve, Nigeria. *Journal of Applied Ecology*, **2**, 367–381.

Hudson, N. W. (1957). The design of field experiments on erosion. *Journal of Agricultural Engineering Research*, **2**, 56–65.

Hudson, N. W. (1971). *Soil Conservation*. Batsford, London.

Israelsen, O. W. and Hansen, V. E. (1962). *Irrigation Principles and Practices* (third edition). Wiley, New York.

Janzen, D. H., Miller, G. A., Hackforth-Jones, J., Pond, C. M., Hooper, K. and Janos, D. P. (1976). Two Costa Rican bat generated seed shadows of *Andira inermis* (Leguminosae). *Ecology*, **57**, 1068–1075.

Jackson, I. J. (1977). *Climate, Water and Agriculture in the Tropics*. Longman, London.

Kalmus, H. (1958). *Simple Experiments with Insects*. Heinemann, London.

Klomp, H. (1966). The dynamics of a field population of the Pine Looper *Bupalus piniarius* L. (Lep., Geom.). *Advances in Ecological Research*, **3**, 207–305.

Kowal, J. M. and Knabe, D. T. (1972). *An Agroclimatological Atlas of the Northern States of Nigeria*. Ahmadu Bello University Press, Zaria.

Lock, J. M. (1972). The effects of hippopotamus grazing on grasslands. *Journal of Ecology*, **60**, 445–467.

Meteorological Office (1956). *Handbook of Meteorological Instruments. Part 1. Instruments for Surface Observations*. HMSO, London.

Monteith, J. L. (1972). *Survey of Instruments for Micrometeorology*. IBP Handbook, **22**. Blackwell Scientific Publications, London.

Odum, H. T., Drewry, G. and Kline, J. R. (1970). Climate at El Verde, 1963–1966. In: *A Tropical Rain Forest*, H. T. Odum (Ed.), pp. B 347–B 418. US Atomic Energy Commission, Oak Ridge.

Penman, H. L. (1963). *Vegetation and Hydrology*. Technical Communication of the Commonwealth Bureau of Soils, **53**.

Pereira, H. C. (1951). A cylindrical gypsum block for moisture studies in deep soils. *Journal of Soil Science*, **2**, 212–223.

Pereira, H. C. (1973). *Land Use and Water Resources*. Cambridge University Press, London.

Peterson, A. (1964). *Entomological Techniques* (tenth edition). Edwards Brothers, Ann Arbor.

Pratt, D. J. and Gwynne, M. D. (1977). *Rangeland Management and Ecology in East Africa*. Hodder and Stoughton, London.

Rapp, A., Berry, L. and Temple, P. (Eds) (1973). Studies of soil erosion and sedimentation in Tanzania. *Geografiska Annaler*, **54A**, 105–379.
Rees, W. A. (1974). Preliminary studies into bush utilization by cattle in Zambia. *Journal of Applied Ecology*, **11**, 207–214.
Schulz, J. P. (1960). *Ecological Studies on Rain Forest in Northern Suriname*. North-Holland, Amsterdam.
Schulze, E. D. and Koch, W. (1971). Measurement of primary production with cuvettes. In: *Productivity of Forest Ecosystems*, P. Duvigneaud (Ed.), pp. 141–157. UNESCO Ecology and Conservation Series, **4**.
Southwood, T. R. E. (1966). *Ecological Methods*. Methuen, London.
Spinage, C. A. (1968). A quantitative study of the daily activity of the Uganda Defassa Waterbuck. *East African Wildlife Journal*, **6**, 89–93.
Trapnell, G. C. (1959). Ecological results of woodland burning experiments in Northern Rhodesia. *Journal of Ecology*, **47**, 129–168.
Walter, H. (1963). Climatic diagrams as a means to comprehend the various climatic types for ecological and agricultural purposes. In: *The Water Relations of Plants*, A. J. Rutter and F. H. Whitehead (Eds). British Ecological Society Symposium, **3**, Blackwell Scientific Publications, London.
Whitmore, T. C. and Wong, Y. K. (1959). Patterns of sunfleck and shade light in tropical rain forest. *Malayan Forester*, **22**, 50–62.
World Meteorological Organization (1968). *Practical Soil Moisture Problems in Agriculture*. World Meteorological Organization Technical Note, **97**.

Circular slide meters for calculating fire-risk are produced by the Australian CSIRO Division of Forest Research, in Canberra.

Index